Changing Lives Through
GENETIC ENGINEERING

Other titles in *The Tech Effect* series include:

Changing Lives Through Artificial Intelligence
Changing Lives Through Robotics
Changing Lives Through Self-Driving Cars
Changing Lives Through 3-D Printing
Changing Lives Through Virtual Reality

The Tech
EFFECT

Changing Lives Through
GENETIC ENGINEERING

Toney Allman

Metropolitan Library System

San Diego, CA

© 2021 ReferencePoint Press, Inc.
Printed in the United States

For more information, contact:
ReferencePoint Press, Inc.
PO Box 27779
San Diego, CA 92198
www.ReferencePointPress.com

ALL RIGHTS RESERVED.
No part of this work covered by the copyright hereon may be reproduced or used in any form or by any means—graphic, electronic, or mechanical, including photocopying, recording, taping, web distribution, or information storage retrieval systems—without the written permission of the publisher.

LIBRARY OF CONGRESS CATALOGING-IN-PUBLICATION DATA

Names: Allman, Toney, author.
Title: Changing lives through genetic engineering / by Toney Allman.
Description: San Diego, CA : ReferencePoint Press, Inc., 2021. | Series: The tech effect | Includes bibliographical references and index.
Identifiers: LCCN 2020016178 (print) | LCCN 2020016179 (ebook) | ISBN 9781682828410 (library binding) | ISBN 9781682828427 (ebook)
Subjects: LCSH: Genetic engineering.
Classification: LCC QH442 .A45 2021 (print) | LCC QH442 (ebook) | DDC 576.5--dc23
LC record available at https://lccn.loc.gov/2020016178
LC ebook record available at https://lccn.loc.gov/2020016179

CONTENTS

Introduction   6
Changed Forever

Chapter One   10
What Is Genetic Engineering?

Chapter Two   21
Drugs and Vaccines for Fighting Disease

Chapter Three   33
Engineering Food Plants for People

Chapter Four   44
Food Animals and Genetic Engineering

Chapter Five   55
Altering Human Genes

Source Notes   67
For Further Research   71
Index   73
Picture Credits   79
About the Author   80

INTRODUCTION

Changed Forever

Brenden Whittaker was born with a rare genetic disorder called chronic granulomatous disease (CGD). It meant that the white blood cells called neutrophils, a critical component of his immune system, were malfunctioning. Over and over, through his childhood and youth, Whittaker suffered serious infections that required hospitalizations and threatened his life. Although he took antibiotics daily, his infections could not be prevented, and granulomas developed throughout his body. Granulomas are lumps of infected tissue that develop as the body tries to wall off infections it cannot fight off. At times, these lumps may obstruct the body's organs and have to be surgically removed, along with surrounding healthy tissue. By the time Whittaker was a young adult, in 2015, he had had a lobe of his liver and half a lung removed, and still he was getting sicker. Half of all patients with CGD die before the age of forty, and Whittaker did not think he had that much time left. He remembers, "I didn't have any kind of plan for the future. I didn't think I was going to need one. I was more focused on the 'right then'—getting healthy right then. . . . I'd never thought that long-term before because I'd been sick for so long."[1]

immune system
The complex network of cells, tissues, and organs that help the body defend against disease and infection

A Trailblazer

Then in December 2015, Whittaker became the first person in the world to take part in a clinical trial of a revolutionary treatment for CGD. At the Dana-Farber/Boston Children's Cancer and Blood Disorders Center, doctors removed a sample of blood stem cells from his bone marrow. In the bone marrow, blood stem cells become the various kinds of blood cells needed by the body, then multiply. Whittaker's stem cells were sent to a lab, where a harmless virus was used to insert the gene for correctly manufacturing neutrophils into the stem cells. The stem cells were multiplied in the lab and then transfused back into Whittaker's bone marrow. The medical researchers hoped that their genetic engineering experiment would boost Whittaker's immune system. Their goal was for the genetically correct stem cells to continue to multiply, making functional neutrophils. They believed that creating even 10 percent of a normal level of neutrophils would prevent constant serious infections from developing. And it worked.

By 2019, Whittaker was a healthy twenty-five-year-old, attending college and planning to continue on to medical school. He has had an occasional bout with a cold or the flu but has recovered as easily as a person without CGD. Doctors have determined that 50 percent of his neutrophils are functioning normally. No one knows whether the changes introduced into Whittaker's body will be permanent. He will be followed and checked medically for fifteen years before researchers know whether the treatment has ended his CGD permanently. Whittaker, however, is looking forward to the future. One of his doctors calls Whittaker's treatment and improvement "a miracle of modern medicine" and says that "it would appear that he is cured."[2]

> **gene**
> A specific DNA sequence that is the basic unit of inheritance and contains the coded instructions for individual traits or characteristics

The Book of Life

This is the hope and promise of genetic engineering for all people born with genetic diseases—that a lifetime of treatments and medicines can be replaced with a one-time cure. Not only sick people, however, can benefit from genetic engineering. Every living thing carries genetic information in the nucleus of almost every cell in the body. Scientists say that genes are like the words in the giant book of life written by deoxyribonucleic acid (DNA). This book makes reproduction possible, whether for bacteria or plants or human beings. It determines how living things develop, what traits and characteristics they have, and what species they are.

Quite often, the book of life contains errors, called variations or mutations, which crept in during reproduction. A genetic mutation is why Whittaker was born with CGD, but it is also why people have different eye colors and blood types, why cherry tomatoes are smaller than standard tomatoes, and why dog breeders can breed poodles of several different colors.

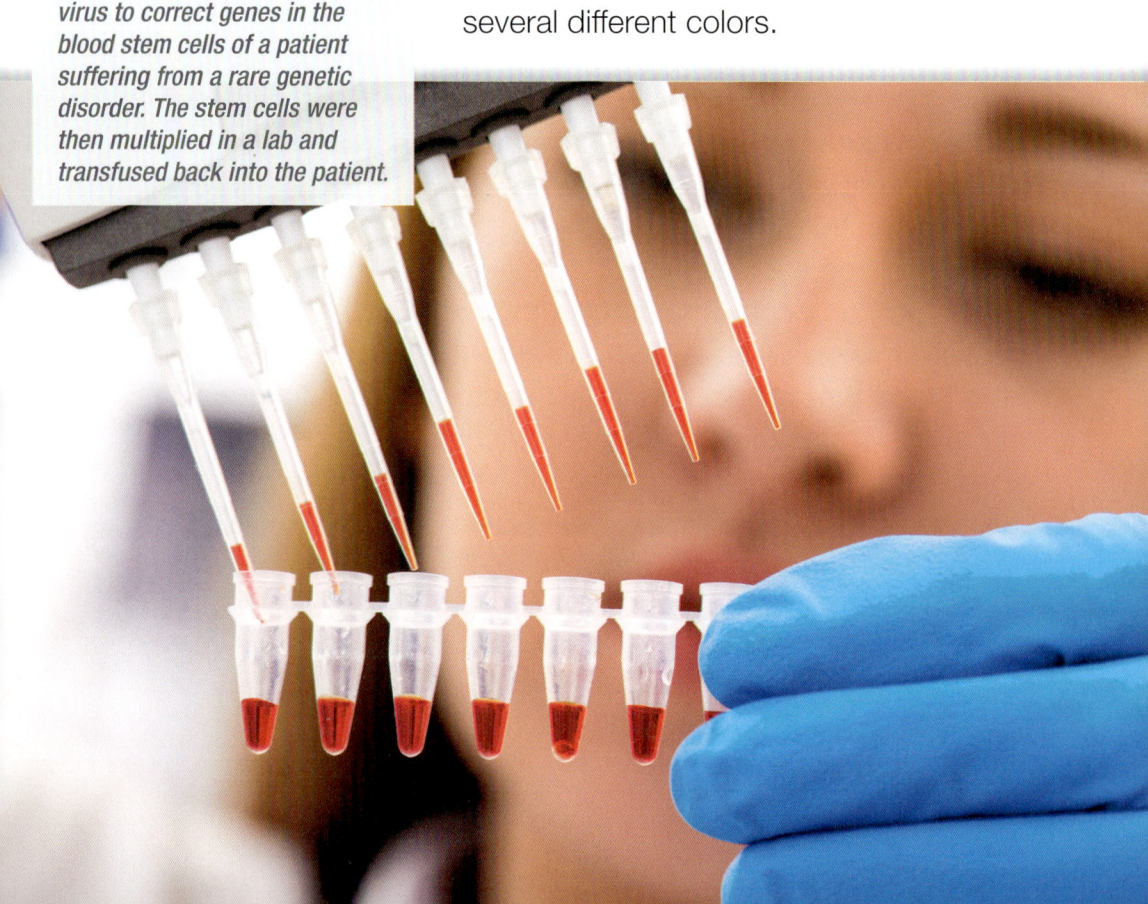

In 2015 doctors used a harmless virus to correct genes in the blood stem cells of a patient suffering from a rare genetic disorder. The stem cells were then multiplied in a lab and transfused back into the patient.

Under Human Control

Until very recently, the genetic book of life was written by nature and was unchangeable by humans, except very indirectly. Now, with today's genetic engineering tools, it is fast becoming possible to precisely and accurately edit the genes of any living thing. The technology could revolutionize the way people live. As science writer Bryan Walsh says, genetic engineering technologies

> herald an era when the book of life will be not just readable, but rewritable. Food crops, endangered animals, even the human body itself—all will eventually be programmable. The benefits are easy to imagine: more sustainable crops; cures for terminal genetic disorders; even an end to infertility. . . . These new technologies offer control over the code of life.[3]

That control and the subsequent impact on society have already begun.

CHAPTER ONE

What Is Genetic Engineering?

In the United Kingdom, a company called Tropic Biosciences is developing a new kind of decaffeinated coffee. About 12 percent of the coffee drunk around the world is decaffeinated, often for health reasons or because people are overly sensitive to the effects of caffeine. The beverage is traditionally decaffeinated by soaking the coffee beans and then using slow cooking and chemicals to leach out the caffeine, but many people do not like the resulting taste. Tropic Biosciences has a better idea. It does not decaffeinate beans. Instead, the company is growing coffee plants that are genetically engineered to produce beans with little or no caffeine in them. The company's CEO, Gilad Gershon, explains, "If you grow the beans without the caffeine or with a lower amount of caffeine to begin with, then you can achieve an end product that is a lot closer in taste to normal coffee, and you can maintain a larger content of the very healthy compounds that are naturally found in coffee."[4]

Genetic engineering, sometimes referred to as genetic modification, means altering the DNA of an organism to change its characteristics. This is what Tropic Biosciences is doing with the coffee plants it is growing in its labs. It is altering their DNA to develop a plant that does not produce caffeine.

Genetics and DNA
All living things or organisms carry DNA in the nuclei of their cells. DNA is a molecule made up of four chemical bases,

referred to as the four chemical letters of the DNA alphabet. The letters are A (which stands for the chemical adenine), T (for the chemical thymine), C (for cytosine), and G (guanine). The bases pair up with each other on the two long strands of the winding ladder of DNA called the double helix. At each step of the ladder, A pairs only with T, and C only with G. The sequence, or order, of these base pairs along the DNA ladder determines the information that builds and maintains each living thing. It is as if the DNA is writing a code and the chemical letters are spelling out the words in a giant book.

Genes are like the words in the book. They are specific lengths or segments of DNA that constitute the basic units of heredity. Because the DNA double helix can make copies of itself, genes are passed on to offspring, half from the male and half from the female. Genes carry the chemical code for making specific proteins.

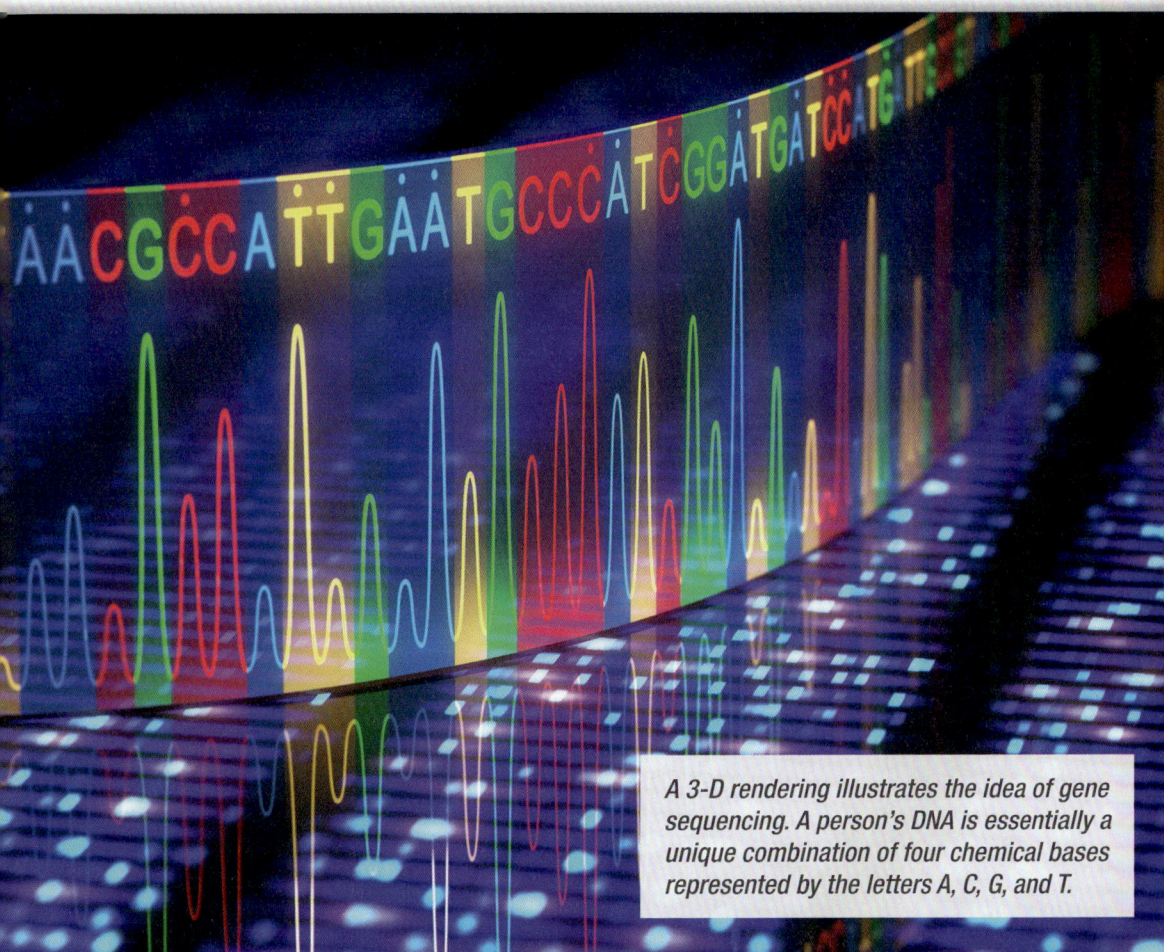

A 3-D rendering illustrates the idea of gene sequencing. A person's DNA is essentially a unique combination of four chemical bases represented by the letters A, C, G, and T.

Proteins do the work of the cells and determine how an organism functions. The genes also determine traits. They determine whether an organism is a bacterium, a mouse, a coffee plant, or a human being. They also code for whether that organism has caffeine in its beans or has blue or brown eyes.

> **chromosome**
> An organized package of DNA in the nucleus of the cell, occurring in different numbers in different organisms

As an example, humans have about twenty thousand to twenty-five thousand functioning genes, all determining traits. These genes may vary in size from just a few hundred to more than 2 million base pairs. Groups of genes are carried in structures called chromosomes. Humans have forty-six chromosomes, arranged in pairs inside the cells.

Inheritance

Chromosomes are inherited from parents. In a microorganism, such as a bacterium, the parent cell is the single cell that reproduces by making two identical copies of the DNA in its chromosome and then dividing into two daughter cells. Whenever DNA is copied like this, there is a chance of errors occurring, and these errors, which are like the misprints and typographical errors in a book, are called mutations. Some mutations are harmful, some make no difference, and some can be beneficial.

In more complex, multicellular organisms, such as humans, sexual reproduction is necessary to create new life. Each parent produces a special cell called a gamete. Gametes are able to split the number of their chromosomes in half when they divide. Thus, when the male and female gametes unite, the resulting offspring has the full number of chromosomes, half from the father and half from the mother. As in bacterial cell division, however, things can go wrong. Errors and mutations can creep in during

> **gamete**
> The male or female reproductive cell that carries half the genetic material of an organism

this process of splitting and recombining of the DNA. On average, each human individual is born with about sixty errors in his or her DNA, most of which are harmless.

Cracking the Code of Life

The entire set of genes in an organism's chromosomes makes up its genome. Scientists have mapped, or sequenced, the genomes of many different organisms, from viruses—which are simply packages of incomplete genetic material—to human beings. According to the National Human Genome Research Institute of the National Institutes of Health (NIH), "The sequence tells scientists the kind of genetic information that is carried in a particular DNA segment. For example, scientists can use sequence information to determine which stretches of DNA contain genes and which stretches carry regulatory instructions, turning genes on or off. In addition, and importantly, sequence data can highlight changes in a gene that may cause disease."[5]

> **genome**
> All the genetic material of an organism

A complete genome sequence for any organism identifies the locations of all the genes and the distances between them in a chromosome. It works out the numbers of base pairs within each gene and the numbers of base pairs between the genes, which are DNA pairs that do not code for traits. Genes that are within the same chromosome are said to be linked. This is important because genes that are close together are likely to be inherited together.

Determining a complete genome is complex because so many base pairs are involved. For instance, the rod-shaped bacterium named *Haemophilus influenzae* has 1.8 million bases in its one chromosome. A round worm—the first animal ever to have its genome mapped—has 100 million bases in six pairs of chromosomes. Thale cress, the first plant to be sequenced, has 119 million bases in five chromosomes. A mouse has 5.4 billion bases in forty chromosomes. And a human has 6.4 billion in forty-six chromosomes. Determining what exact lengths of DNA sequences make up a gene is a hugely complex undertaking and has not been completely accomplished. Many organisms, however, have had some of their genes specifically identified.

In humans, almost the entire DNA is completely sequenced, but all of the genes have not yet been determined. Scientists do know that 99.9 percent of the genetic makeup of all humans is identical. The differences in the remaining 0.1 percent are what account for the uniqueness of every individual. The same is true for animals. Some of the genes or variations in genes that code for these differences are known, and some are not.

Even when all the genes in an organism are identified correctly, it does not mean that scientists know what every gene does or which ones are linked or whether mutations in specific genes affect functioning. Nevertheless, says science writer Edd Gent, "The ability to read genomes has transformed

> **Scientists know that 99.9 percent of the genetic makeup of all humans is identical. The differences in the remaining 0.1 percent are what account for the uniqueness of every individual.**

our understanding of biology."[6] A significant result of this transformation is that it allows scientists to use genetic engineering to edit and modify the genome within the DNA sequences where the gene function is known. Researchers do not need to know the complete genetic makeup of any living thing to be able to change its DNA and alter one or more of its traits. They just need to know the location and function of the gene they want to modify.

Recombinant DNA Technology

Altering the genome of an organism can be done in different ways. Recombinant DNA technology is the process by which scientists take one piece of DNA and combine it with another strand of DNA. Generally, it involves joining the DNA molecules of two different species. This joining produces new genetic combinations in an organism. It is accomplished by a process analogous to the word processing functions of cutting, pasting, and copying. Chemically, scientists cut a DNA strand, paste in a new piece of DNA, and then insert it into another organism. There, the new genetic instructions code for a new function.

This process of joining two pieces of DNA was made possible by the discovery of restriction enzymes in bacteria. Bacteria are often attacked or infected by viruses. These viruses inject their genetic material into the bacteria, forcing the bacteria to incorporate the virus's genetic instructions and thus reproduce the virus. Eventually, so many viruses are made that the bacteria burst, killing them. The new viruses then move on to infect more and more bacteria.

Many bacteria, however, developed a defense. They carry restriction enzymes that find the alien viral genetic material and chop it into pieces. Different strains of bacteria produce different restriction enzymes. Scientists have identified more than four thousand restriction enzymes that can be used as a kind of molecular scissors to cut specific DNA sequences. The restriction enzyme known as *Eco*R1, for instance, cuts the DNA string GAATTC. Wherever this string of DNA appears, the restriction enzyme cuts it at a specific point.

Today, scientists can use restriction enzymes to cut the DNA of bacteria or other microorganisms and then insert new, foreign DNA sequences from another species—animal, plant, or human. The sequences, which may be a gene, part of a gene, or multiple genes, are glued into place with another enzyme called a ligase. The resultant altered molecule can replicate itself like any DNA molecule, producing millions of altered cells. As an example, 94 percent of the cheese in the United States is produced using recombinant DNA technology. The enzyme chymosin is required for making cheese. It is naturally found in rennet, a complex of digestive enzymes found in the stomachs of calves. Formerly, rennet was collected when the calves were slaughtered for meat. Now, instead of slaughtering the calves to get the rennet to make cheese, the gene that codes for making chymosin is snipped from a DNA sample taken from a calf's stomach. Then it is pasted into a suitable microorganism, one that is able to produce chymosin, and it becomes a little chymosin factory. Science writers Jon Entine and XiaoZhi Lim explain further: "These genetically modified microbes are allowed to multiply and [are] cultivated in a fermentation process while they produce and release chymosin into the culture liquid. The chymosin can then be separated and purified."[7] The resulting large quantities of the enzyme are used for cheese production.

Gene Editing

Recombinant DNA technology has been a remarkable scientific success story, but a new genetic engineering method gives scientists an even more powerful ability to easily modify genes and DNA. It is called gene editing, or CRISPR/Cas9 (CRISPR, for short). Gene editing is precise, easy, and inexpensive for scientists to accomplish and does not involve inserting genes from one species into another. It is a way of intentionally creating a mutation in a species' genetic material—a mutation that is desirable and purposeful.

Gene editing with the CRISPR/Cas9 system was invented in 2011 by Jennifer Doudna of the University of California–Berkeley

Prime Editing

Still in the research stage of development is a gene-editing technology that is even more precise than CRISPR/Cas9; it is called prime editing. Occasionally, CRISPR/Cas9 changes genes that it should not. This happens because it cuts through both strands of the DNA double helix. Then the technology depends on the cell's own repair mechanisms to knit the strands back together. The repair does not always work as expected because base pairs can be added or deleted during the repair process. Such changes are called "off-target mutations."

Prime editing avoids this problem. The method involves altering the Cas9 protein so that it just cuts one DNA strand instead of cutting through both strands. Then an RNA guide molecule called pegRNA is added. RNA is only a single strand, instead of a double helix, but it is composed of chemical matches to DNA sequences. The pegRNA carries instructions for making a new DNA sequence. It has an enzyme added to it, called a reverse transcriptase enzyme, to accomplish this. The enzyme translates the RNA instructions to make a new DNA strand from the RNA and inserts it at the cut site. Prime editing was developed in 2019. It has only been used on cells in a lab as of mid-2020, but if it works in higher organisms such as animals, scientists in the near future may have an easy, safe, and perfectly precise way to repair malfunctioning genes.

and Emmanuelle Charpentier of Umeå University in Sweden. The researchers were studying an unusual repeating cluster of DNA sequences in some bacteria. The clusters are known as Clustered Regularly Interspaced Short Palindromic Repeats (CRISPR). Scientists already knew that these sequences were part of bacteria's immune systems for fighting off viruses. The bacteria store bits of chopped up viral genetic material in their CRISPR spaces as a kind of memory bank incorporated into the bacteria's own genomes. If the same virus attacks again, the bacteria recognize it immediately and produce attacking enzymes called Cas9. Cas9 enzymes search for viral ribonucleic acid, or RNA, matches. Although RNA is similar to DNA, DNA carries the genetic code,

Gene editing with the CRISPR/Cas9 system was invented in 2011 by Jennifer Doudna (pictured) of the University of California–Berkeley and Emmanuelle Charpentier of Umeå University in Sweden.

while RNA reads the code and actually makes the proteins for cell functioning. When Cas9 recognizes an RNA match from a previous infection, the enzyme cuts out that exact sequence from the virus's genetic material, thus rendering it harmless.

Doudna and Charpentier discovered that they could fool the Cas9 enzyme. They could introduce an artificial RNA sequence into the CRISPR sequence, and the Cas9 enzyme would search for anything with that code (not just from viruses) and start cutting. In 2012, the two researchers wrote a scientific paper describing how they could use CRISPR/Cas9 to cut any genome in any living thing at any specific sequence they chose. In 2013, Feng Zhang of the Broad Institute of Harvard and the Massachusetts Institute of Technology (MIT) demonstrated that he could precisely edit mouse and human cells in lab dishes using CRISPR/Cas9. Another scientist at Harvard edited human cells in his laboratory at the same time. There was no longer a need for restriction enzymes. Scientists no longer had to merge the DNA from one species of organism with that of another. Doudna says, "It was really quite amazing how quickly it was possible to harness this technology once it was clear how it operated. . . . Amazingly, this technology operates efficiently in virtually all cell types of organisms in which it's been tested."[8]

Genetic Engineering Revolution

In the twenty-first century, CRISPR has become the kind of genetic engineering that scientists believe could change the world. Researchers used to have to spend thousands of dollars and weeks of time altering a gene. With CRISPR, it takes them just a few hours and costs about seventy-five dollars. With the technology, they can "silence," or cut out, any gene in any living thing. In the laboratory, this means a fast, easy way to compare how different organisms function with and without that gene and thus

How a Mutation Can Cause Disease

The four letters of the DNA alphabet form the words of the genetic code, which are always written and read in groups of three. These three-letter words, or triplets, provide all the instructions needed to make proteins. In English, instead of in chemical letters, for example, a DNA string of instructions might look like this: THEFATCATATETHEBIGRAT. Broken into groups of three, the sentence makes perfect sense: THE FAT CAT ATE THE BIG RAT. However, if the letter *C* is dropped from the sentence, it would be analogous to a DNA mistake occurring as a cell is duplicating its DNA and preparing to divide—one tiny piece is omitted from the code. Now, the string of letters looks like this: THEFATATATETHEBIGRAT. Broken into words of three letters each, it reads like this: THE FAT ATA TET HEB IGR AT; it becomes gibberish and makes no sense. The same thing happens when a genetic mutation takes place. The instructions no longer make sense. That one small error can cause a serious disease in which the chemical directions in the body do not work. Mutations may involve a missing letter, an added letter, or a substituted letter, but in all cases, they may be important enough to create a genetic disease.

learn the trait the gene codes for. They can try to cut out genes in plants, animals, and people that cause problems or diseases, and they may be able to add a new gene to take the place of one that was cut out. Doudna says,

> CRISPR-Cas9 is a revolutionary, once-in-a-generation tool that offers the real potential to quickly and efficiently achieve what was once thought impossible. Since 2012, the technology has been adopted rapidly, transforming basic research, drug development, diagnostics and agriculture. . . . As we move into [the 2020s], it is clear that CRISPR-based applications will help us tackle societal challenges including disease, food production and environmental sustainability."[9]

Much of the gene editing being done today is still in the research stages, but CRISPR technologies hold great promise and are rapidly moving from the laboratory into the real world. Research is also continuing on advanced uses for recombinant DNA technology, as gene editing cannot replace this technology altogether. No matter what methods are used, however, genetic engineering is giving humans control over the very code of life.

CHAPTER TWO

Drugs and Vaccines for Fighting Disease

In South Africa, one thousand people a day contract HIV. HIV/AIDS continues to be a scourge and major killer in many parts of the world, particularly in sub-Saharan Africa. Researchers funded by the National Institute of Allergy and Infectious Diseases (NIAID) in the United States began working on a new vaccine for the disease in 2016. Using recombinant DNA technology, the researchers developed two vaccines, to be administered together.

Vaccines protect against diseases by presenting the immune system with microorganisms that have been rendered harmless but still build immunity. In the NIAID HIV vaccine study, fifty-four hundred sexually active South African volunteers participated in a trial of the new vaccines' effectiveness. The researchers did not expect a 100 percent success rate because HIV mutates into many different subtypes so frequently. Even partial success, however, could make a big difference. Dr. Anthony Fauci, NIAID's director, said at the study's onset, "Even a moderately effective vaccine would significantly decrease the burden of HIV disease over time in countries and populations with high rates of HIV infection."[10]

Learning from Mistakes

Despite the researchers' high hopes, however, on February 3, 2020, NIAID announced that it was halting the trial. The vaccine was a failure. Half the volunteers in the study had received the vaccine. The other half had received a placebo injection containing no vaccine. In the group receiving the vaccine, 129 ended up infected with HIV. In the group receiving the placebo, 121 developed HIV infections. These results demonstrated basically no difference in infection rates among the two groups. Glenda Gray, one of the study's directors in South Africa, reported, "Years of work went into this. It's a huge disappointment."[11]

No vaccine technology is guaranteed to work every time, but researchers are not giving up. HIV/AIDS is an epidemic in South Africa. Researchers wonder whether the infected volunteers were exposed to too many subtypes of the virus and too repeatedly during the four years of the trial. Or perhaps the scientists chose the wrong genes or not enough of them. Perhaps, some others hypothesize, constant exposure played a role in the vaccine's failure. Fauci speculated that perhaps the vaccines were not strong enough and were overwhelmed by the virus. Still, AIDS experts like Mitchell Warren, the executive director of the AIDS Vaccine Advocacy Coalition, insist that genetically engineering an effective HIV vaccine is possible, but as Warren notes, "We also know that HIV is the most challenging virus for which we've ever tried to create a vaccine."[12] Genetic engineering is a complex process, even when the technology is well understood, but recombinant DNA technology has worked for many other vaccines and medicines. Someday, researchers believe, it will work for HIV/AIDS, too.

Recombinant DNA Technology and Insulin

In actuality, recombinant DNA technology has changed the face of medicine and medical research. In a multitude of ways, altering genes in bacteria and viruses has benefited people. It all began with the first genetic engineering success in 1982, with the development of synthetic insulin for people with diabetes. The pancreas (a digestive organ in the abdomen) of a diabetic person does not produce any or enough of the hormone insulin. With-

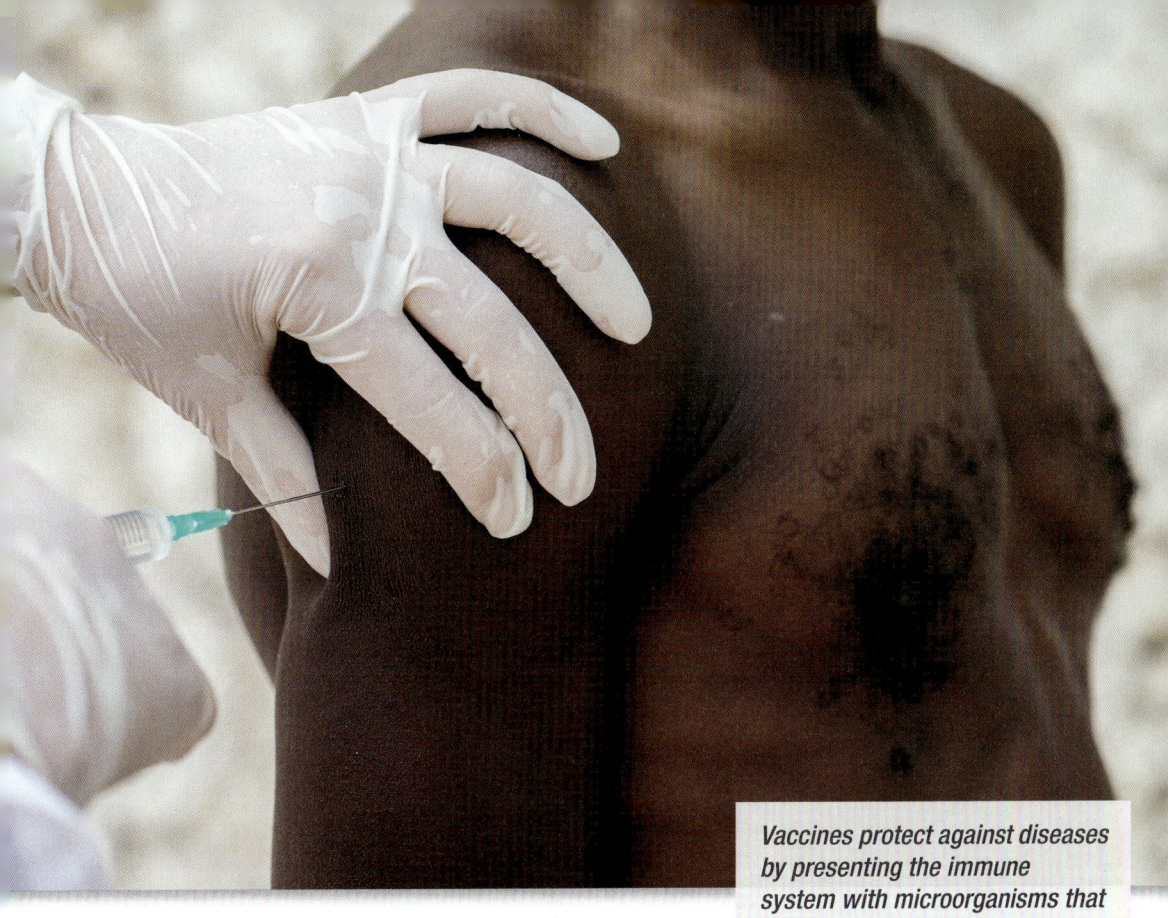

Vaccines protect against diseases by presenting the immune system with microorganisms that have been rendered harmless but still build immunity. In a National Institute of Allergy and Infectious Diseases HIV vaccine study, fifty-four hundred sexually active South African volunteers participated in a trial of the new vaccines' effectiveness.

out insulin, the sugar in the bloodstream from ingested food cannot get inside the body's cells to feed them, so without insulin, people will die. In the past, insulin was extracted from the pancreases of slaughtered cattle and pigs, but recombinant DNA technology changed that.

Herb Boyer and Stanley Cohen were genetic experts who believed they could produce a better insulin for diabetics. The human genome had not yet been sequenced, but scientists did know the gene that codes for insulin production. Boyer and Cohen worked out a method of recombining that human gene with bacterial genetic material and then using the bacteria to produce human insulin. Bacteria are single-celled microorganisms with a nucleus inside the cell with genes that carry DNA instructions, just as human cells do. Bacteria, however, have structures that human cells do not.

These structures are called plasmids and are small loops of DNA outside the nucleus that are passed on to future generations along with the DNA in the nuclei. Because they are outside the nuclei, plasmids are easier to remove from the bacterium, alter, and then reinsert. Bacteria also reproduce very rapidly through cell division, with one bacterium dividing into two identical daughter cells. The daughter cells can replicate themselves in as little as twenty minutes, meaning that a colony of bacteria can double in size every twenty minutes. Given an ideal laboratory environment and enough food, one bacterium could become 1 billion in ten hours. Cohen and Boyer knew that if they could change the DNA in the plasmids of just one bacterium, they would have millions of insulin-producing bacteria to meet human needs.

plasmid
A circular strand of DNA outside the nucleus of a bacterium

The two scientists chose a harmless strain of E. coli bacterium that is found in the human digestive system to use as an insulin-making factory. They removed a plasmid from the bacterium and cut it with restriction enzymes. The plasmid was ready for the introduction of the human gene that codes for making insulin, but there was a problem. The human gene also contained noncoding DNA known as introns, as do all human genes. Bacteria have no introns, just as humans have no plasmids, so the bacterial RNA could not read the human DNA codes. To solve the problem, Boyer and Cohen had to make copies just of the parts of the gene that code for insulin, which they accomplished using chemicals in their lab. This gave them DNA sequences that the bacterium's RNA could read. Now the human insulin-making DNA code could be inserted into the plasmid loop, which was then glued back together with ligase. Finally, the recombined plasmid was

introns
Noncoding sequences of DNA within a gene

A Vaccine for COVID-19

During the 2020 pandemic of the respiratory disease COVID-19, researchers and pharmaceutical companies around the world addressed the challenge of vaccine development against the new disease. Since vaccines normally take about ten years to develop, this was no easy task. Nevertheless, one company called Moderna Therapeutics, working with the NIAID, genetically engineered a possible vaccine.

The virus that causes COVID-19 is an RNA virus, named SARS-CoV-2. Instead of DNA, its genetic sequences are packages of single-stranded RNA. Like developing a DNA vaccine, Moderna scientists created a vaccine using mRNA (messenger RNA). In humans, information from DNA is transferred to mRNA which is then read by the cell to produce a protein. Researchers developed a synthetic mRNA sequence and combined it with the code for the spikes on the shell of the coronavirus. Those spikes are used by the virus to break into a cell. The spikes give the virus its crown-like appearance from which it gets its name. They are antigens for the immune system but do not carry the genetic codes within the shell that cause sickness. When injected under the skin of a person, the mRNA sequence enters the cells, begins instructing the cells to make "crown points" and potentially triggers an immune system response to the foreign instructions.

By mid-May, 2020, early trials of Moderna's vaccine in a few volunteers had already demonstrated safety and the production of antibodies. This was a hopeful sign and trials were continuing.

---

reinserted into the bacterium, which was allowed to divide and multiply into a whole colony of bacteria, each of which carried the insulin-making gene.

The two scientists carried out the same process with many bacteria and ended up with huge colonies of genetically engineered bacteria, all producing insulin. All that was left to do was harvest the "crop." It was real, human insulin, not insulin processed from farm animals. It did not cause any allergic reactions in people, as animal insulin sometimes did. It was easily absorbed by human bodies and worked more quickly than animal insulin did as well. It was the first drug created using recombinant DNA technology.

Modern Recombinant DNA Drugs

Today, almost all the insulin used in the United States and in much of the world is synthetic insulin manufactured with recombinant DNA technology. Bacteria or other simple organisms, such as yeast cells, are used to produce it. The technology has been improved and simplified; different forms of insulin, such as fast-acting versus slow-acting, have been developed, but the basic technology is much the same. The insulin is as pure and perfect as the insulin produced by the human body.

Recombinant DNA technology has improved the lives of millions of people with diabetes, but it was just the beginning of the genetic engineering of medicines. Variations of the technology are used to produce many valuable medicines today. Human growth hormone (HGH), for example, is produced using the plasmids in bacteria. HGH is used to treat children born with some forms of dwarfism. These children's pituitary glands do not produce enough growth hormone to grow to a normal height because the gene that codes for the hormone is defective. If they are treated with genetically engineered HGH early enough, however, these children can still attain a normal height. People with hemophilia can also be treated with medicine made from recombinant DNA technology. Hemophilia is a blood disorder caused by a genetic error that reduces the ability of the blood to clot. For a hemophiliac, a small cut that would clot in seconds in a person without hemophilia can cause a dangerous loss of blood. The gene coding for the clotting factors is known, and those clotting factors can be synthetically produced with recombinant DNA technology in the same way that insulin is. Other medicines produced through recombinant DNA technology include drugs for treating the inflammation of arthritis and the reddened skin flaking of psoriasis, as well as drugs that increase the human immune system's ability to fight some kinds of cancers.

Reproducing Insulin with Bacteria

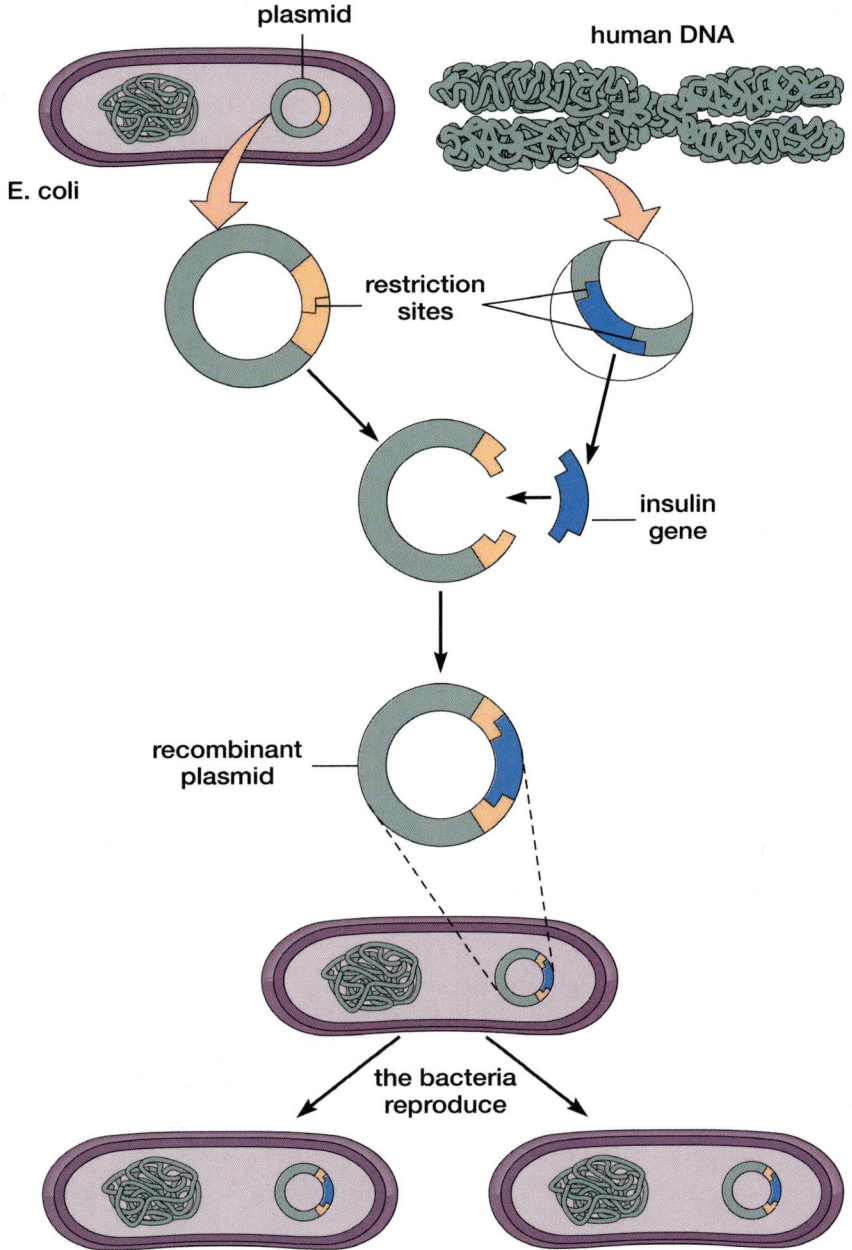

E. coli bacterial cells carrying copies of the insulin

A plasmid from a harmless E. Coli bacterium is removed and recombined with an insulin gene from human DNA. Then the recombinant plasmid is reinserted into the bacterium, which will divide and multiply into a whole colony of bacteria, each of which carries the insulin-making gene.

Harnessing the Immune System and Vaccines

The immune system is the body's defense network that fights off infections and other invasions of the body. When it detects an alien invader such as a bacterium or a virus, the immune system is responding to a chemical substance or substances on that invader. The immune system begins producing antibodies specific to that substance. Any substance that provokes an antibody response is called an antigen. The antibodies attach to the antigen and mark the invader for destruction by other immune system cells.

Once the infection is conquered, the antibodies are retained by the immune system. Should that same pathogen (disease-causing organism) attack again, the immune system can now recognize it and respond faster than the first time, before it causes sickness. Thus, the body is now immune to the disease. Vaccines mimic this situation by introducing harmless pathogens into the body. These will not cause disease but will still trigger the production of antibodies so that the body will be ready if a future encounter with a dangerous version should occur. Genetic engineering is leading to vaccines that are safer and more effective than ever before.

Traditionally, vaccines have been made with killed or weakened bacteria or viruses that produce immunity without causing the disease. There are some disadvantages, however, to this approach, especially with viruses. Live viruses, for instance, will make some people sick because they do not have a strong immune system. Others are allergic to the chicken eggs in which the

> **antigen**
> Any substance that stimulates an immune system response

> **antibody**
> A protein made by immune system B cells in response to an antigen; each antibody binds to only one specific antigen and marks it for destruction by other immune system cells

weakened viruses are grown. Also, these vaccines do not provide long-lasting protection and require regular booster shots. Another disadvantage, notes research scientist Kris Heeter, is that "in some instances . . . the impaired virus can revert back to an active virus and cause the illness it was designed to fight." Genetic engineering, however, provides a solution, as Heeter explains: "Modern advances in genetics and recombinant DNA, or rDNA, technology have enabled scientists to create vaccines that no longer have the potential to cause disease."[13]

Making Recombinant DNA Vaccines

Viruses cause sickness by hijacking a body cell and forcing it to use the virus's genetic instructions to replicate more viruses. Each virus has a surface coating made of proteins. Some of these are the antigens that provoke an antibody response, though they do not themselves cause illness. Vaccine developers can isolate the genetic instructions for such surface proteins, recombine them with the DNA of other microorganisms, and replicate them in the lab. The result is a vaccine that can stimulate the production of antibodies but that cannot possibly cause illness.

The first recombinant vaccine developed in this way was for hepatitis B. It combines a hepatitis B virus surface protein with yeast cells to produce an effective, safe vaccine. A recombinant influenza vaccine was first approved in the United States in 2013. To make this vaccine, scientists identified the DNA code for making a flu virus's surface protein known as hemagglutinin (HA). HA is the antigen that triggers the human immune system to make antibodies against that flu virus. That antigen is combined with a virus that can infect only invertebrates (animals without skeletons). The recombined viruses rapidly reproduce the HA antigen. The Centers for Disease Control and Prevention (CDC) explain further: "This antigen is grown in bulk, collected, purified, and then packaged as recombinant flu vaccine."[14] Other recombinant DNA vaccines in use today include a vaccine for meningitis B and one for human papillomavirus (HPV).

The Next Generation: DNA Vaccines

To use genetic engineering to develop any vaccine, scientists must know which genetic instructions code for antigens in a pathogen, and many pathogens are still poorly understood. Other pathogens, such as HIV, attack the immune system itself, making it extremely difficult to train the immune system antibodies to recognize the threat. With others, as the previously mentioned failed HIV vaccine trial suggests, scientists have not yet determined which proteins on the pathogen's shell are antigens that will trigger strong antibody production. Scientists have sequenced the genomes of more than twelve hundred viruses and continue to sequence more. Still, says Heeter, "despite all the vaccines developed through rDNA technology, infectious diseases in animals and humans continue to be a worldwide problem."[15]

Part of the answer to this problem may lie in what is known as the third generation of vaccines, the first generation being the one using killed or weakened pathogens, and the second the one using

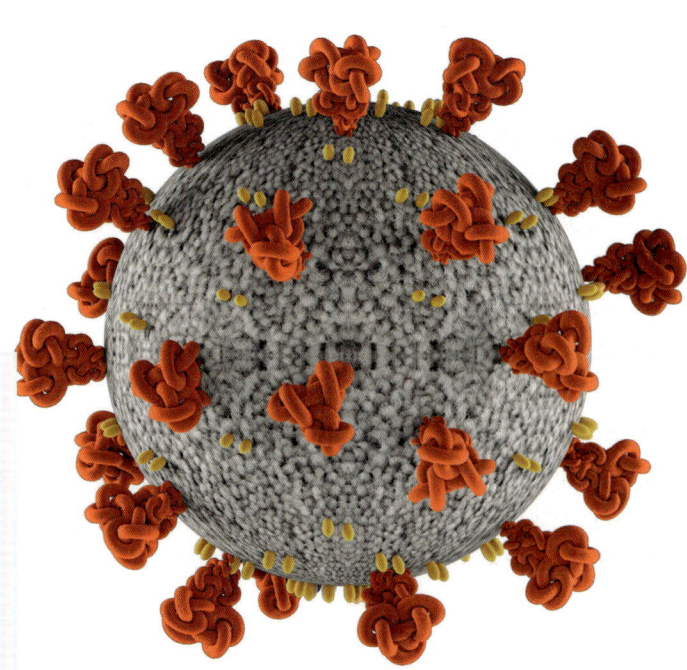

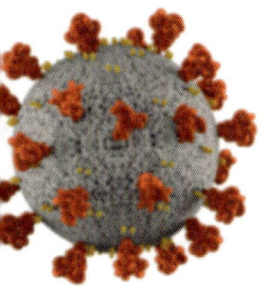

A 3-D rendering of the COVID-19 virus shows surface proteins (red and yellow) that are unique to all viruses. Vaccine developers isolate these surface proteins, recombine them with the DNA of other microorganisms, and replicate them in the lab.

Toward a Zika Virus Vaccine

Zika is a mild flu-like disease caused by a virus spread through mosquito bites. Outbreaks of the disease have occurred in Africa, the Americas, Asia, and the Pacific Islands. Although the disease in adults is mild, it has devastating results for unborn children. Zika-infected pregnant women give birth to babies with severe birth defects, such as microcephaly (small head size), brain damage, and other physical malformations. No vaccine yet exists to counter Zika disease.

In December 2019, a team of researchers at the University of Adelaide in Australia reported a breakthrough in the search for a Zika vaccine. The team developed a DNA vaccine using Zika virus sequences that include the coding for making a protein that allows the virus to replicate. The team injected the vaccine into mice in their laboratory and then exposed the mice to the actual virus to see whether the vaccine worked. It did. The T cells in the immune systems of the mice attacked and killed the virus. The researchers call their vaccine a T cell–based vaccine, because it triggers T cell attacks rather than triggering antibody production. The next step in development is to try the vaccine in a phase I trial with people. Lead researcher Branka Grubor-Bauk says, "If we can progress this work and immunise women who are of reproductive age and most at risk, we can stop the devastating effects of Zika infection in pregnancy and make a huge difference to the health of the global community."

Quoted in University of Adelaide, "Breakthrough in Zika Virus Vaccine," ScienceDaily, December 13, 2019. www.sciencedaily.com.

---

recombinant technology. The third generation is DNA vaccines. Instead of growing antigen proteins to produce a vaccine, scientists are working on developing vaccines using pieces of viral DNA that are directly injected into the body under the skin. Just as happens with actual virus infections, a few of the body cells take up the viral instructions and begin producing the antigens from the virus. The white blood cells of the immune system, specifically the antibody-generating B cells and the pathogen-destroying T cells, recognize the antigens as invaders, and a strong immune response is stimulated. This is called cell-mediated immunity, and scientists believe it might work even with diseases for which standard vaccines fail.

Hope for the Future

DNA vaccines are still in the research stage of development. A major problem is getting body cells to actually take in the viral DNA. Researchers today are embedding the code in bacterial plasmids and then using short bursts of electrical current applied to the skin to create small pores in the cells where the DNA carrying plasmids are injected. This allows the viral DNA to enter some of the cells. In animals, the technique has produced immune system responses to the vaccines for some infections. In 2019, for example, the United States Department of Agriculture (USDA) approved the first DNA vaccine for the H5N1 influenza virus in chickens. The vaccine seems to provide full immunity for chickens against the disease, although it may require booster shots. Vaccine researchers Seyed Davoud Jazayeri and Chit Laa Poh describe the advantages of this DNA vaccine: "DNA vaccination provided a new and valuable approach to the development of poultry vaccines and offered advantages in flexibility of design, speed, simplicity of production, and the ability to elicit . . . immune responses."[16] They and other veterinary researchers hope that this vaccine is but the first of many to come for protecting poultry and other animals from serious diseases, which often can be transmitted to humans as well.

No DNA vaccines are yet available or approved for humans, but tests in animals infected with human diseases such as HIV, malaria, influenza, and rabies have shown that immune responses do occur. In January 2020, the results of a phase I clinical trial of a DNA vaccine for treating cancer was published. (A phase I clinical trial tests the safety of a vaccine in human volunteers.) The vaccine was determined to be safe. In the phase II trial, researchers will test how well the vaccine works to teach the immune system to attack and destroy cancer cells.

DNA vaccines for people may not be a reality yet, but rapid progress is being made toward success. Researchers expect DNA vaccines to be faster and cheaper to produce than other vaccines and to be completely safe. Already, standard vaccines and recombinant vaccines have prevented suffering in and saved the lives of millions of people. In the near future, DNA vaccines may prevent the diseases no other vaccines have yet been able to.

CHAPTER THREE

Engineering Food Plants for People

Approximately 75 percent of the world's usable land today is devoted to agriculture, and according to the United Nations Intergovernmental Panel on Climate Change, that land is under tremendous pressure. As the same crops are grown on the same acreage over and over again, the soil loses nutrients. Most agricultural lands are at capacity, meaning that crop yields are as high as they can get. Farmers try to compensate for this with intensive fertilizer and pesticide use, which ultimately damages the environment. Yet the world's population and its demand for more food continue to increase. In addition, climate change may affect which staple crops can grow where and may incapacitate some areas of agricultural lands. Genetic engineering promises to offer a solution to these problems and be the answer to feeding the world's people. Connor McKoy, communications director for the biotechnology trade association Bio, comments that "though far from a cure-all, the potential for gene editing to make every acre of land more productive even in the face of climate change has captured the imagination of plant scientists, the agtech industry, and governments alike."[17]

Gene editing is the use of CRISPR technology to modify the genetic information in a plant or other living thing to change its traits. Many scientists believe that gene editing can create food plants that will grow in poor soil, plants that need fewer pesticides to protect them from insect damage, and increased crop yields no matter what the climate.

Despite fears about agricultural lands being at capacity, these scientists believe that abundant crops will provide ample food for all. Yiping Qi, a plant geneticist and engineer at the University of Maryland, says, "It's hard to say for sure what we can achieve in the next few decades, but I think with Crispr we have a chance to catch up to climate change."[18]

Genetic Engineering Before Gene Editing

Gene editing is the newest form of genetic engineering, but before CRISPR technology there was genetic modification, which creates genetically modified organisms (GMOs). Genetic modification, as opposed to gene editing, is the technology of combining the DNA of two different species of organisms using recombinant DNA technology. In the case of a plant, the goal is to create a new, improved kind of plant that would probably never occur naturally. As is done with medicines, viruses or bacteria are used as vectors, or vehicles to carry the new information into the DNA in the nuclei of plant cells. The altered cells are grown into plants in the lab. Seeds produced by these modified plants inherit the new DNA.

Bt corn, for example, is corn genetically engineered to be resistant to attacks from caterpillars such as the European corn borer. It kills the caterpillars that eat the corn plant. Bt is short for *Bacillus thuringiensis*, a bacterium that lives in the soil and carries a protein that is toxic to many forms of caterpillars that attack crops. The Bt gene that codes for the toxic protein has been identified, pasted into a plasmid, and then inserted into the corn's genetic material. As the corn plant grows, its cells produce the toxin because they now carry the code for making it. Any worm that feeds on the plant will die.

The advantages of Bt corn are undeniable. Standard insecticides must be sprayed over entire fields of corn and often miss many corn borers. Corn borers, as their name implies, tunnel into

> **vector**
> A vehicle, such as a virus or plasmid, that carries a foreign piece of DNA into a cell

Approximately 75 percent of the world's usable land today is devoted to agriculture. As the same crops are grown on the same acreage over and over again, the soil loses nutrients.

the cornstalk where they are hidden from insecticide sprays. No protection exists, however, when the very cornstalk itself is toxic. Most other insects, including beneficial ones, are not harmed by Bt corn toxins. Farmers are not exposed to harmful chemicals because they do not have to spray for pests. Because there are no holes from borers in the stalks, secondary pests such as fungi and other insects cannot enter the stalks and infect or destroy the plants. The environment is safe from pesticides because there is no need for farmers to use them. Crop yields are higher with Bt corn than with non-engineered corn, so more corn is available for the market, which means higher profits for the farmer and lower prices for the consumer. Today, more than 90 percent of the corn crop in the United States is genetically engineered.

The GMO Controversy

Many people worry that genetically modified food plants like Bt corn are unsafe for the environment or will somehow affect other

Saving Chocolate with CRISPR

Because of rising temperatures and reduced rainfall from climate change, scientists in the country of Ghana, the world's second-largest producer of cacao, from which cocoa and chocolate are made, predict that their crop could be extinct by 2080. In fact, some scientists predict that the cacao trees could become extinct globally in forty years. Researchers in Ghana and elsewhere are looking to gene editing in hopes of developing a drought- and heat-resistant cacao tree. At Pennsylvania State University, for instance, molecular biologist Mark Guiltinan is experimenting with CRISPR to delete the TcNPR3 gene in cacao beans. With this gene deleted, the trees have increased resistance to diseases and also grow fast. Since diseases and insect pests easily attack drought-stressed plants, his research may help cacao trees survive changing climate conditions.

Under hotter conditions, cacao trees are especially susceptible to cocoa swollen shoot virus, a disease that first originated in Ghana and has spread throughout several African countries. Millions of trees are lost to the disease each year. Guiltinan is searching for the gene in cacao's DNA that controls the response to cocoa swollen shoot virus disease and has identified several promising genes. He says, "I see a strong possibility of the first gene-edited cacao being ready for farmers in about five to 10 years."

Quoted in Joseph Opoku Gakpo, "Gene Editing Could Save Ghana's Cocoa from Extinction, Scientists Say," Cornell Alliance for Science, June 13, 2019. https://allianceforscience.cornell.edu.

---

food crops grown nearby. A 2018 study from University of Maryland researchers demonstrated the opposite. Bt corn seems to benefit other crops grown in nearby fields, as well as the farmers who grow those crops. The study's researchers reported that "Bt corn suppresses pests regionally, with declines expanding beyond the planted Bt crops into other non-Bt crop fields. We show that widespread Bt field corn adoption is strongly associated with marked decreases in the number of recommended insecticidal applications, insecticides applied, and damage to vegetable crops in the United States."[19]

Genetically modified foods have also given rise to concerns that they are possibly unsafe for humans to eat. Some people claim that such foods may cause allergic reactions in vulnerable people or cause cancer by introducing new and unnatural DNA sequences to the body. There is no direct evidence for these concerns. According to the American Association for the Advancement of Science (AAAS),

> The World Health Organization, the American Medical Association, the U.S. National Academy of Sciences, the British Royal Society, and every other respected organization that has examined the evidence has come to the same conclusion: Consuming foods containing ingredients derived from GM (genetically modified) crops is no riskier than consuming the same foods containing ingredients from crop plants modified by conventional plant improvement techniques.[20]

GM Foods Around the World

Although many developed countries restrict or even ban GMOs, the United States approves their use after extensive trials and testing for safety. Today, more than 90 percent of US-grown corn, soybeans, and rapeseed (from which canola oil is made) is genetically modified, as is 100 percent of sugar beets. Some of these crops are altered to resist pests whereas others are made to be herbicide resistant, so that farmers can spray fields for weeds without harming the crop. Most scientists agree that genetically modified food crops can dramatically increase the amount of high-quality food available globally.

The public often rejects GMOs, but such foods are impossible to avoid altogether. Foods like soybeans and corn are ingredients in many foods that people eat, such as cereal, snack foods, and vegetable oils used in cooking. The Food and Drug Administration (FDA) states on its website, "Since GMO foods were introduced in the 1990s, research has shown that they are just as safe as non-GMO foods."[21] Nevertheless, by an act of Congress, the USDA has begun implementing a regulation requiring manufacturers to disclose on food packaging labels whether certain foods have been bioengineered. "Bioengineered foods" means GM foods and are defined as those that have been "modified through certain lab techniques and cannot be created through conventional breeding or found in nature."[22] Starting on January 1, 2020, large manufacturers had to disclose these ingredients on some food labels, such as for pink-fleshed pineapple, virus-resistant papaya, some canola oil, and corn. All such foods must be labeled by all marketers by 2022.

No one knows whether consumers will be disturbed by the "Bioengineered" label, and many ingredients in prepared foods or at restaurants are exempt from the labeling anyway. In Vermont, however, where bioengineered foods are already labeled

bioengineered
In food products, an ingredient containing genetic material that has been modified by recombinant DNA technology

As is done with medicines, viruses or bacteria are used to carry new information into the DNA of plant cells. The altered cells are grown into plants in the lab.

as such, one 2019 University of Vermont study found that opposition to GM food fell by 19 percent after the state's labeling law went into effect. Disclosure of GM foods might make people feel more comfortable about the technology. Most scientists and agricultural experts hope that is the case. Globally, GMO crops are already changing lives for the better, especially in developing countries. Small farmers and their customers in poorer countries are increasing wealth and health because of increased yields and an increased and more stable food supply. Paul S. Teng, plant researcher and board chairman of the International Service for the Acquisition of Agri-biotech Applications (ISAAA), asserts, "Biotech crops offer enormous benefits to the environment, health of humans and animals, and contributions to the improvement of socioeconomic conditions of farmers and the public."[23]

The CRISPR Difference

The biggest reason that GMOs cause so much controversy is that they are recombinant crops, combining the DNA of two distinct organisms such as would never happen in nature. Genetic engineering using CRISPR technology, however, is very different. This gene-editing technology introduces no foreign genes into a plant that could not occur naturally. It involves only altering a specific gene or DNA sequence in the plant's genome, much as could happen naturally during normal reproduction. CRISPR researchers believe the technology will revolutionize food production in the near future.

For the most part, CRISPR gene editing is still confined to research labs and greenhouses, but its use with food plants is advancing rapidly. Jennifer Doudna, coinventor of CRISPR technology, says, "I think in the next five years the most profound thing we'll see in terms of CRISPR's effects on people's everyday lives will be in the agricultural sector."[24] In the United States, unlike in the European Union, gene-edited foods are not subject to regulation as GMOs are. The USDA ruled in 2018 that it will not regulate such crops if the genetic changes could have been produced through conventional plant breeding. Thus, nothing stands in the way of CRISPR plant breeding, except perhaps for public acceptance.

Gene-Editing Accomplishments

Public perception is why Yinong Yang, a professor at Pennsylvania State University, has voluntarily applied for FDA approval for his gene-edited white button mushrooms, which do not turn brown as they age. He explains, "Before taking it to the market, it is essential to demonstrate scientifically that a gene-edited crop is as safe as conventional and organic crops."[25] Yang and his research team are interested in extending the shelf life of foods and helping to reduce the tremendous food waste that occurs around the world. In the United States, according to the FDA, about 30–40 percent of the food supply—some 133 billion pounds— is wasted each year. This discarded food could instead have been used to feed hungry people. Yang's mushrooms are a first small step in changing that situation.

Mushrooms turn brown as they begin to decompose due to the production of enzymes called PPOs that are produced by vegetables and fruits as they age. Mimicking a natural genetic mutation, Yang and his team used CRISPR/Cas9 technology to precisely snip out one of the six genes that code for the production of PPOs in white button mushrooms. This small genetic change reduced the tendency of white button mushrooms to turn brown by 30 percent, thus extending the shelf life and home refrigerator life of the mushrooms. The mushrooms are not yet for sale, but Yang reports he has eaten them himself in his lab.

Food waste reduction is just one of the areas in which CRISPR technology could improve the food supply. Doudna explains, "There's lots of research being done to alter plant properties that will allow plants to resist drought, to resist disease, potentially to be more nutritious, and to do that using gene editing so that genes can be very precisely altered without requiring years of breeding." She goes on to describe researcher Zach Lippman and his gene-editing work with tomatoes at Cold Spring Harbor

Researchers used CRISPR/Cas9 technology to precisely snip out one of the six genes that code for the production of PPO enzymes in white button mushrooms. This small genetic change extended the shelf life of the mushrooms.

Laboratory in New York. He is using CRISPR/Cas9 technology to control the number of fruits a tomato plant can produce. Doudna says of this discovery, "So you could start imagining being able to control crop production in many different kinds of crops using this sort of a strategy, which sounds very exciting."[26] Plants with increased fruits could make a major difference in the crop yields available to feed the world.

More Research Needed

Editing plants to resist diseases is somewhat harder to accomplish. In China, plant researcher Gao Caixia is working with wheat. She has identified six genes in wheat that make it susceptible to powdery mildew, a fungal disease that can wipe out wheat crops. Although she and her team could delete those genes easily with CRISPR technology, they found they were also removing desirable traits. They have begun trying to add new wheat genes for resistance, but adding is difficult compared to deleting with CRISPR. Gao says, "We are not so good at it."[27] Nevertheless,

Fortified Bananas

For thirteen years, scientists at Uganda's National Agricultural Research Laboratories (NaRL) worked to develop healthier cooking bananas (also known as plantains), and by 2018, they declared that they had succeeded. The cooking bananas the researchers developed are enriched with vitamin A, and each banana contains half the vitamin A content needed daily by children and pregnant women. This is important because cooking bananas are a staple food in the diets of Ugandans, but are very low in vitamin A. About 28 percent of preschool children in Uganda are deficient in the vitamin, as are 23 percent of pregnant women. Vitamin A deficiency can cause blindness, stunted growth, reduced resistance to infections, and even death.

The scientists at NaRL identified genes in nonedible bananas from Southeast Asia that code for high vitamin A production. They inserted these genes into two local varieties of cooking bananas and grew the plants. Every banana produced had a pinkish color due to the higher vitamin A content. Dr. Priva Namanya Bwesigye, the laboratory head, notes, "We have also ensured that the banana plant grows true to the traditional type with the initial traits remaining intact, apart from introduction of Vitamin A gene." The NaRL has applied to the Ugandan government to approve the genetically modified banana plants and expects to release them for field trials by 2021.

Quoted in Lominda Afedraru, "Uganda's Researchers Ready to Taste Test Their GMO Vitamin A–Enriched Banana," Genetic Literacy Project, March 12, 2018. https://geneticliteracyproject.org.

---

the science is young, and researchers such as Gao expect success as they practice and learn.

Other genetic researchers are struggling to delete the genes in foods like peanuts and wheat that cause allergies or disease in people. While theoretically possible, the work is not easy. For example, people with celiac disease cannot digest the protein gluten in wheat, but to make wheat edible for them, scientists would have to snip out forty-five genes. So far, they have succeeded with only thirty-five genes. With peanuts, only thirteen genes code for the allergens in people, but those genes also code for most of the healthy proteins in the peanuts. A healthy, safe peanut is a real gene-editing challenge. Gene editing is not perfect, but scientists still expect many future benefits.

The Future Is Now

The first gene-editing success to reach the market in the United States is an improved soybean oil. It is not yet available in stores, but as of 2019, it was being used in some restaurants in the Midwest. The oil, called Calyno, is made from soybeans whose genes have been edited to be more nutritious and heart healthy than regular soybeans. Calyno has less saturated fat than other soybean oil and a longer shelf life. The shelf life is important because most soybean oil has to be hydrogenated to make it last for a long time and not spoil. The downside of hydrogenation, however, is that it produces unhealthy trans fats. So avoiding the need for hydrogenation is another plus for Calyno.

Calyxt, the company that produces Calyno, is working to develop and produce several gene-edited food plants, as are many other companies around the world. Dan Voytas, the chief science officer at Calyxt, says, "Right now the food industry solves all its problems through processing or chemistry. We'd like to do it through genetics and gene-editing."[28] That includes research on such crops as wheat with more fiber and less gluten, potatoes that can tolerate cold storage without breaking down, and herbicide-resistant rapeseed for canola oil.

On its website, Calyxt says, "We're committed to making a difference by developing stronger plants that will improve health and sustainability for people and the planet."[29] Gene editing may be the genetic engineering breakthrough that protects humanity's future and ensures everyone is fed.

> **hydrogenated**
> Having hydrogen added in order to turn a liquid oil into a solid oil, which makes it more stable and so increases its shelf life

CHAPTER FOUR

Food Animals and Genetic Engineering

On March 8, 2019, the FDA approved the breeding of unique Atlantic salmon eggs on a fish farm in Indiana. The facility is owned by a company called AquaBounty Technologies, and it is hatching and breeding the eggs in large tanks. After eighteen months of growth, the fish will be ready to be sold for food in the United States. The salmon, known as AquAdvantage salmon, have been gene edited to grow twice as fast as regular salmon, while needing 25 percent less food. This means they reach marketable size faster and at a lower production cost, thus increasing profits for the company while increasing the available food supply for consumers.

At the same time, claims AquaBounty Technologies, the genetically engineered salmon are helping to save the environment. Less feed for the salmon means less agricultural land needed for growing their food. Also, wild Atlantic salmon have been overfished for so long that they are now on the endangered species list and can no longer be caught and sold. Atlantic salmon eaten in the United States have to be farm raised, which has proved a difficult and expensive operation. AquAdvantage salmon could change that. AquAdvantage salmon have been approved and sold in Canada since 2017 and are expected to be sold in the United States by the end of 2020.

AquAdvantage salmon is the world's first—and so far only—gene-edited food animal approved for human consumption. Most researchers, however, are convinced that genetic engineering has the potential to provide an abundance of high-quality animal protein to feed the world's burgeoning population. Millions of people in developing countries are emerging from conditions of poverty and are demanding access to more animal foods. Eric Hallerman is a Virginia Polytechnic Institute and State University fisheries scientist and an FDA adviser. He says of AquAdvantage salmon, "People want to eat more meat. We have to do it efficiently. So, I think this has to be part of that."[30]

Obstacles to Gene-Edited Food Animals

Unlike the USDA's decision to exempt gene-edited food plants from regulation, the FDA has ruled that gene editing in food animals must be regulated. In 2017, the FDA explained that "in general, the agency considers genomic alteration to meet the definition of a New Animal Drug, unless otherwise excluded, because it is intended to affect the structure or function of the animal." In other words, the animals themselves are not considered to be drugs but as "containing new animal drugs."[31] The FDA was concerned about the unintended consequences of off-target mutations that could occur from the new breeding methods; that is, that the animal might have a harmful, unexpected trait. The FDA Center for Veterinary Medicine's director, Steven Solomon, explains, "We have a public health obligation to protect consumer and animal health."[32]

This decision, however, means that researchers and companies end up spending millions of dollars and years of time seeking FDA approval for gene-edited food animals, and it has slowed further development of those animals in the United States. Some companies and researchers are moving to less-regulated countries to continue their research and development. AquaBounty Technologies for instance, moved the production and marketing of its new gene-edited tilapia to Argentina, where

Wild Atlantic salmon have been overfished for so long that they are now on the endangered species list. Atlantic salmon eaten in the United States now have to be farm raised.

the government labeled them as a new breed of animal rather than a genetically modified one.

Many researchers and biotechnology companies criticize the FDA's decision. Randall Prather, for example, a leading genetic engineer at the University of Missouri, says that regulating animals altered with CRISPR technology as if they were GMOs "just doesn't make sense."[33] He explains that the changes made with gene editing cannot be distinguished from the natural mutations that occur spontaneously in all living things. Researchers are not giving up, but for now, much of the work in genetic engineering in animals is taking place outside the United States.

Progress Toward Improved Livestock

China's government is hugely supportive of CRISPR research and the gene editing of animals to potentially help feed its 1.4 billion people. Jennifer Doudna says, "This is a country and a culture that really values science and technology. Their government has put very serious money into it, and they're walking the walk."[34] Two gene-editing researchers who are taking advantage of this government support are Wang Haoyi and Zhou Qi of the Chinese Academy of Sciences Institute of Zoology in Beijing. Wang and Zhou used CRISPR technology to make a genetic change in pigs that speeds up their growth, altering a gene that codes for a growth hormone. In their laboratory, the two scientists used CRISPR technology to edit the genes of pig embryos and then implanted those embryos inside female pigs' wombs. The piglets that were born not only grew faster themselves but also passed on the gene changes to their own offspring. Someday, such research could give rise to whole new breeds of pigs that grow quickly and reach market weight earlier than other breeds of pigs, meaning more meat for everyone.

A different approach to boosting livestock production is being researched by a US company called Acceligen. Scientists there are working on developing heat-resistant cattle. The goal is not to enable more cattle to be bred on the planet but to develop healthier, more productive cattle. Cattle usually do not cope well with heat. The best breeds for milk production and producing meat are subject to heat stress and physical damage or death in the wrong kinds of climates. In tropical Africa, for instance, a typical dairy cow produces ten times less milk than a cow in temperate Europe. Acceligen researchers identified a gene in an unusual cattle breed in the Virgin Islands that makes the animals naturally heat resistant. They are very poor milk producers, but they have a thin hair coat and sweat easily. Researchers named the gene for these heat tolerance traits the "slick gene."

In the laboratory, Acceligen scientists inserted the slick gene into the embryo of a Red Angus cattle breed, and the altered embryo was surgically transplanted into a Red Angus cow. The resulting calf, named Genselle, was born in 2018 in Minnesota and then moved to Brazil, where she is growing normally. Tad Sonstegard, chief scientific officer at Acceligen, says, "She acts like a normal animal with no signs of heat stress in what is now the middle of summer in Brazil. And that is very unusual."[35] Researchers are still studying Genselle, ensuring that she grows and behaves like any other dairy cow, and they hope she will be as good a milk producer as any other Red Angus dairy cow.

Helping the Poorest People

Professor Appolinaire Djikeng believes that heat-resistant animals could be a boon to poor farmers in Africa. He is the director of the Centre for Tropical Livestock Genetics and Health in Edinburgh, Scotland, and is working toward a future in which gene editing lifts millions of Africans out of poverty. Djikeng is working in collaboration with Acceligen on the problem of heat tolerance and also researching ways to prevent animal diseases. In many countries in Africa, people are subsistence farmers, owning perhaps only one or two animals and depending on them completely for their livelihoods. They need sturdy farm animals, yet often face the tragedy of an animal dying or producing so poorly that the farmer and family face starvation. As a native of the African country of Cameroon, Djikeng remembers, "Growing up, I understood that if you are farming and you are that vulnerable, there has to be something there to help, perhaps resilient animals, disease-resistant animals, and developing the best practices. At the time, the science was not good enough to make a difference. And it was my commitment to change that. It was a personal mission."[36]

> **subsistence farming**
> Farming in which crops and animals are grown to meet the survival needs of the farmer and his or her family, with very little or no surplus that can be sold

Djikeng is fulfilling that mission with an effort to prevent some of the most devastating animal diseases using gene editing. At his center, funded by the Bill and Melinda Gates Foundation, he and his scientific team are working to develop chickens that are resistant to

In many countries in Africa, people are subsistence farmers, owning perhaps only one or two animals and depending on them completely for their livelihoods.

Newcastle disease, a lethal viral disease that can wipe out whole flocks. The scientists are researching the genomes of chickens in order to understand which DNA sequences may code for disease protection. They know from experience that African chicken breeds tend to be more resistant to many diseases than the breeds in the Western world. African breeds, however, are also low in egg production and do not grow as well or as big as other breeds.

As of 2019, Djikeng's team had analyzed the genomes of 580 birds of sixty different breeds, searching for the genes for disease resistance. Some African chickens, for example, catch Newcastle disease and die, but others do not, and the team wants to know why. They can compare the DNA sequences in the genomes, looking for variations. Once likely gene candidates are found in the strongest chickens, the researchers can delete those genes

Editing Mosquitoes

People do not eat mosquitoes, but mosquitoes are important when it comes to food animals and people. Mosquitoes transmit many diseases including malaria in people and West Nile Virus in both animals and people. Some scientists are researching ways to eradicate these diseases by targeting the mosquitoes that carry them rather than the diseases themselves. At Imperial College London in the United Kingdom, for instance, researchers led by biologist Andrea Crisanti are targeting malaria. The scientists used CRISPR technology to break and damage the X chromosome in the sperm of a small population of male mosquitoes. Since females need two X chromosomes, this gene mutation meant that mostly male mosquitoes were hatched when the males mated with normal females. Male mosquitoes do not bite and cannot spread disease. When a small population of these mosquitoes was placed in cages with hundreds of other normal mosquitoes, the entire mosquito population died out because quite soon no females were produced. The same technique might someday wipe out mosquito-borne diseases in food animals because there would be no more mosquitoes. The question is, according to scientists, whether humans should wipe out whole species of insects and whether the results for the environment would be beneficial or disastrous.

with CRISPR technology and test the chickens for disease resistance. If the chickens get sick, the scientists know that they have identified a disease-resistance gene. Then they will be able to use CRISPR gene editing to insert that gene into healthy, highly productive chickens of any breed. Productive, fast-growing chickens that are disease-resistant could improve the lives of the poorest poultry farmers in Africa.

Using CRISPR to Fight Avian Influenza

Chickens around the world are also vulnerable to avian influenza, which not only wipes out flocks but can also spread to humans and cause serious epidemics. In 2019, researchers at Imperial College London and the University of Edinburgh experimented with gene editing chicken cells in the lab. First, the scientists determined that the influenza virus attacks a specific protein molecule in chicken cells called ANP32A. This turned out to be the key molecule that the virus uses to replicate. Next, the researchers identified the DNA sequence that codes for the production of ANP32A. Finally, they deleted the DNA sequence with CRISPR technology. The altered cells resisted avian influenza attacks because the protein the virus uses to replicate was now gone. Mike McGrew, one of the lead scientists in the research, explains, "This is an important advance that suggests we may be able to use gene-editing techniques to produce chickens that are resistant to bird flu. We haven't produced any birds yet and we need to check if the DNA change has any other effects on the bird cells before we can take this next step."[37]

If the researchers are able to produce healthy gene-edited chickens, they will have found a way to prevent a major disease that threatens chickens worldwide. They may even find the answer to stopping a future human epidemic. Chickens are an animal reservoir for flu viruses, meaning a place that the virus ordinarily lives. When a chicken flu virus mutates to being able to infect humans, it can mean a new kind of flu that spreads easily from person to person. CRISPR technology to produce resistant chickens could make such a scenario impossible. Wendy Barclay, a lead investigator in the study, says, "In this research, we have identified the smallest possible genetic change we can make to chickens that can help to stop the virus from taking hold. This has the potential to stop the next flu pandemic at its source."[38]

reservoir
The animal, plant, person, soil, or any substance in which an infectious organism normally lives and multiplies

CRISPR and Pig Diseases

A deadly disease in pigs called African Swine Fever (ASF) cannot spread to humans, but it has terrible consequences nonetheless. It has a 100 percent mortality rate in infected pigs. Beginning in 2018, an outbreak of ASF killed 25 percent of the world's pig population. China has been especially hard hit. During 2019 alone, 100 million pigs in China were killed by ASF. Chinese consumers eat 66 percent of the world's pork, and that dependency created a crisis during which farmers were bankrupted while the cost of pork doubled. Using CRISPR technology, Chinese researchers are attempting to both rapidly diagnose the disease to help prevent its spread and to develop an ASF-resistant pig.

In 2020, a team of research scientists in China reported using CRISPR to analyze the genome of the virus and identify its presence in a pig's blood sample. The team was able to develop an easy, rapid test for the virus that can be used to detect the infection in pigs in as little as one hour. Other researchers are urgently working to create an ASF-resistant pig. In 2018, two researchers, Lai Liangxue and Ouyang Hongsheng of Jilin University, developed pigs resistant to Classical Swine Fever (CSF), which is similar to ASF but caused by a different virus. The scientists knew that some wild boars are naturally resistant to CSF and were able to insert a resis-

A deadly disease in pigs called African Swine Fever (ASF) wreaked havoc on pig populations. Beginning in 2018, an outbreak of ASF killed 25 percent of the world's pig population.

Pig Organs for People

Organ transplants save lives and relieve suffering, but around the world donated human organs are in short supply. In the United States, for example, more than 112,000 people were awaiting organ transplants in early 2020, and 20 people die every day waiting for an organ transplant. In 2019 in the United Kingdom, 6,077 people were on waiting lists for transplants, and 408 died while waiting. In China in 2019, the organ shortage was critical, with 300,000 people needing organ transplants and only 10,000 organs available.

Chinese biologist Lai Liangxue is using CRISPR technology in an effort to solve the organ shortage by modifying pig organs for human use. He hopes to help people with heart disease, diabetes, failing kidneys, and eye diseases, among others. Human bodies reject animal organs because the immune system recognizes them as foreign and produces antibodies to attack and destroy them. Lai and his research team are attempting to identify and delete the pig genes that cause such rejection. The scientists believe that as many as twenty genes may be involved, and so far, they have succeeded in finding and deleting four. They plan to transplant a gene-edited pig organ into a monkey first and then eventually test such organs in people. If the research team succeeds, xenotransplantation, as it is called, (*xeno* comes from the Greek for "foreign") could end organ shortages and save the lives of millions of people around the world.

tance gene from wild boars into domestic pigs. So far, however, this has not protected pigs from ASF.

Scientists also have determined that the ASF virus infects wild pigs, like warthogs and bushpigs, without making them sick. Researchers are using warthog DNA sequences to edit domestic pigs' genomes. This would not make pigs resistant to ASF, but it might create pigs that do not sicken and die even when infected by the ASF virus, a situation called resilience, as opposed to resistance. A major problem with this gene-editing idea is that

resilience
The ability to cope with an infectious organism without disease symptoms, as opposed to being able to resist infection altogether

many countries would not allow ASF-infected animals to be bred. The fear would be that the animals would become reservoirs for the virus and infect other pig populations, and so research continues.

More Work Needed

In pigs alone, Chinese researchers have successfully made more than forty different modifications with CRISPR technology, but still neither they nor any other researchers have succeeded in introducing a gene-edited animal breed into the farming community or the larger society. Partially, this slow progress is due to government regulations, even in China, that require several years of safety testing before genetically modified animals can be sold commercially. Partially, it is due to fear that public opinion will cause any company selling genetically engineered animals to fail. Jennifer Kuzma, a genetic engineer at North Carolina State University, points out, "All the polls indicate that people are less comfortable with animal biotechnology than plant biotechnology."[39]

In part, this is because gene-editing technology—as exciting as it is—is not perfect. Responsible researchers continue to be concerned about unintended and unexpected changes to DNA that may occur in animals. In 2019, for example, scientists at the FDA discovered just such an error in the genomes of hornless dairy cattle that had been developed using a molecular editing technology similar to CRISPR called TALENS. A fragment of the DNA from the plasmid used to carry the hornless gene into the cattle embryos had been incorporated into the grown cattle's genomes, instead of disintegrating as intended. It was not necessarily a dangerous error, but it demonstrated the necessity for rigor and precision when using gene-editing tools.

Despite the need for further research, geneticists Christine Tait-Burkard and the team at the Roslin Institute at the University of Edinburgh argue that CRISPR technology "can provide huge benefits to the livestock sector and will have a transformational impact on the ability of our species to produce sufficient food in an environmentally sustainable way."[40]

CHAPTER FIVE

Altering Human Genes

When she was six months old, Céline was diagnosed with type 1 spinal muscular atrophy (SMA). SMA is a rare genetic disease caused by a mutation in the gene called SMN1. This gene codes the making of a protein needed for the functioning of the motor nerves (known as neurons) in the spinal cord. The Cleveland Clinic explains, "Without adequate SMN protein, spinal cord motor neurons begin to shrink and die. As this happens, the child's brain is unable to control the body's voluntary muscles, especially those in the arms and legs and in the head and neck. The muscles begin to weaken and waste away. This affects movements such as walking, crawling, head and neck control, swallowing, and breathing."[41] At the age of six months in 2019, Céline already could not lift her head, roll over, or sit up.

Before the advent of genetic engineering, children like Céline died by the age of two. Céline, however, was treated at the Children's Hospital of Philadelphia (CHOP), where she received two new breakthrough gene therapies. Gene therapy is a form of genetic engineering. Céline's first gene therapy was with a drug called Spinraza, which was approved by the FDA in December 2016 and is administered by spinal injection. It works by increasing the production of the SMN protein by other, backup genes.

The First Gene Therapy for SMA

Everyone has two SMN1 genes, one from each parent, that produce most of the SMN protein. Another pair of genes,

SMN2, code for producing about 10 to 15 percent of SMN proteins in the motor nerve cells. Babies with SMA have damaged or missing SMN1 genes, but they still have SMN2 genes. Spinraza therapy delivers an extra RNA sequence that attaches to any SMN2 gene and increases its ability to code for the essential protein. (RNA reads and carries the instructions of DNA to the cell.) Spinraza, however, must be injected regularly because its effects diminish over time. SMN2 genes just do not make stable proteins like SMN1 genes do. Spinraza increases the life span and improves motor function in children with SMA, but it cannot reverse all the damage done to muscles before treatment begins, so the earlier it is given in life the better.

A Second SMA Treatment

Céline began receiving Spinraza therapy and a second gene therapy, called Zolgensma, was added to her treatment. Zolgensma was still undergoing FDA clinical trials at the time, but it was so promising that the FDA gave Céline's doctors special permission to try it. Zolgensma was ultimately FDA approved on May 24, 2019. The therapy is remarkable because it actually replaces the missing SMN1 gene. It is delivered into the spinal fluid as a one-time injection, after which the patient's own body begins to produce SMN proteins.

Céline's doctors and family have high hopes for her. At the time of her first birthday, she was already able to sit up on her own and to stand with leg braces. The doctors believe she will walk someday. Although Zolgensma cannot reverse all motor damage from SMA, many children who receive it in infancy are making normal muscle-growth progress. Céline's neurologist, Elizabeth Kichula, says that

> while SMA remains a serious and life-threatening disorder, Spinraza and Zolgensma are changing the outlook for patients. They can stop progression of [the] disease, making early diagnosis and rapid treatment critical to ensuring the best outcomes. While they are still not cures, they allow for

continued motor improvement, particularly in the setting of supportive families and a commitment to rehabilitative therapy.[42]

Gene therapy is literally life saving for children with SMA. Gene replacement is so new that the long-term outlook is still unknown for children treated with Zolgensma, but it has been administered to children between birth and age two for about four years. All of the recipients are still alive, and none require ventilators to breathe because their muscles are strong enough. Many are able to sit and stand, and some walk, dance, and play outside.

A woman with spinal muscular atrophy (SMA) sits in wheelchair. SMA is a rare genetic disease affecting the spinal cord and caused by a mutation in the gene called SMN1.

Gene Therapy for Transplantation

Gene therapy developments are not limited to children with SMA. Gene therapy is revolutionizing treatments for many genetic diseases, especially those caused by the mutation of a single gene. Gene therapy is generally performed in the same way: A harmless virus is used as a vector to carry new genetic information into a cell to treat a specific disease. Viruses are used because of their ability to slip inside cells, infect them, and take over the genetic instructions. In the case of gene therapy, a harmless virus is used, and the "infection" is a healthy copy of a gene that is missing or mutated. The virus with its cargo of a healthy gene can be introduced into cells in two ways. It may be injected directly into the patient's body, or it may be added to a sample of cells that have been removed from the patient's body and then put back into the patient once they are genetically corrected. Either way, the goal is to create cells that function correctly because they have the correct DNA instructions.

Kymriah, for instance, is a gene therapy treatment for a kind of cancer called acute lymphoblastic leukemia (ALL), which most commonly strikes children and young adults. It is a cancer of the white blood cells known as B cells that are supposed to fight infections. Instead, because of faulty genetic coding, the stem cells in the bone marrow, where all blood cells are made, do not develop correctly and multiply out of control. Out-of-control cell multiplication is the definition of cancer. This was the life-threatening situation that Paulina faced. She had been fighting ALL since 2008 when she was seven years old, but traditional cancer treatments did not work for her. Her cancer would disappear for a while but always returned.

In 2014, doctors at CHOP decided to treat her with Kymriah, a gene therapy performed outside the body. First, doctors collected white blood cells called T cells from Paulina's blood. T cells are the body's killer cells that destroy invaders. In the laboratory, these T cells were infected with a virus that had been gene-edited to code for recognizing the antigens on cancerous B cells. Then the T cells were multiplied in the lab until doctors had millions of them to transfuse back into Paulina's body. There, the genetically

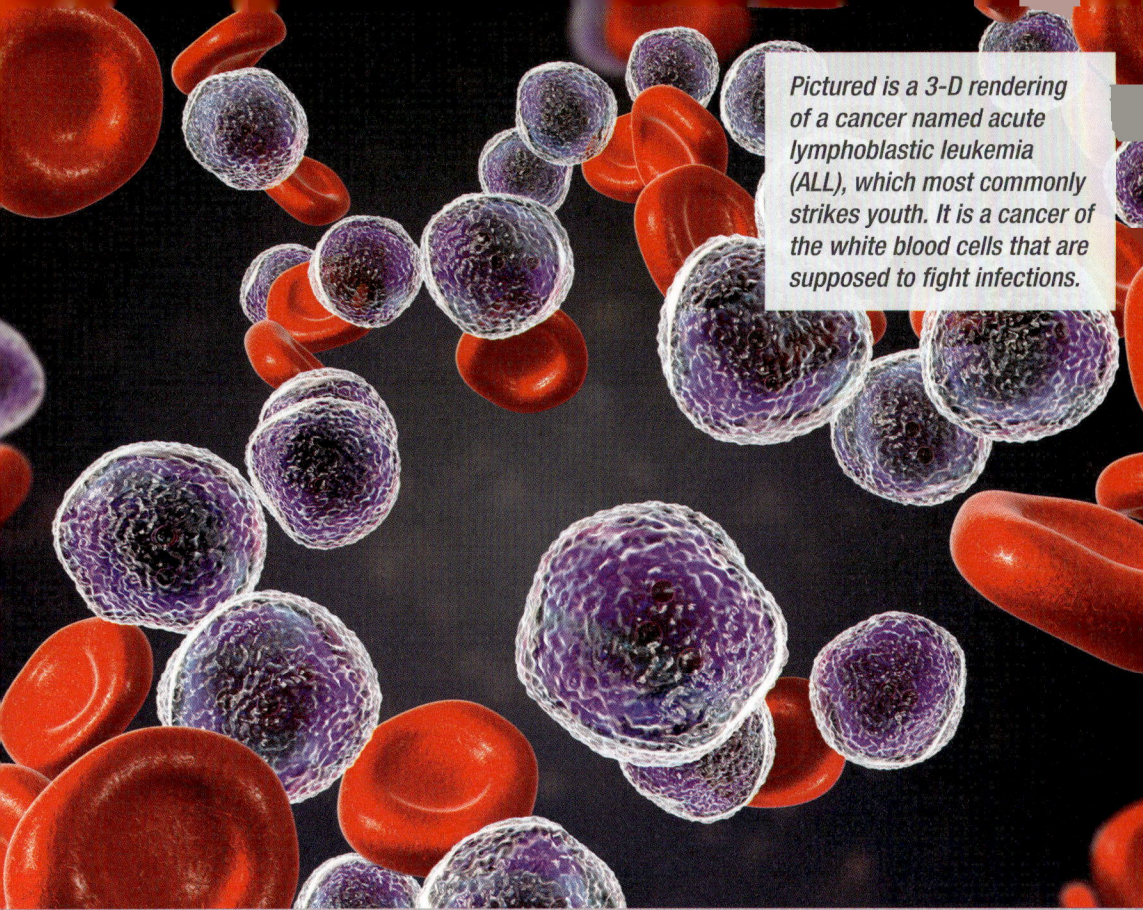

Pictured is a 3-D rendering of a cancer named acute lymphoblastic leukemia (ALL), which most commonly strikes youth. It is a cancer of the white blood cells that are supposed to fight infections.

altered T cells went to work, recognizing and destroying any cancerous B cells produced in her bone marrow. After one injection, Paulina's cancer disappeared, as it has with many other young people with ALL who are treated with Kymriah. By age sixteen, Paulina was still cancer free.

Directly into the Body

Jack Hogan, on the other hand, received his gene therapy directly into his eyes to correct a genetic blinding disease. Jack was born with retinitis pigmentosa, which is caused by a mutation of the RPE65 gene. The gene is missing the DNA sequences that code for making proteins that allow the eye to function correctly. Although he was not completely blind, Jack had no night vision, no peripheral vision, and limited ability to read or do other visual tasks. Doctors predicted that he would be completely blind by age forty. In 2018, when he was thirteen years old, Jack received a gene therapy called

Luxturna at the Massachusetts Eye and Ear Infirmary in Boston. Luxturna uses a viral vector to carry a healthy DNA sequence derived from human cells directly into the retina of the eye. In a delicate surgical procedure, Jack's surgical team carefully injected Luxturna directly under the retina of the boy's left eye. A week later, the surgery was repeated on his right eye. With the healthy RPE65 gene now in his retinal cells, they began to function normally.

retina
The layer of tissue at the back of the eye that senses light and sends signals to the brain through the optic nerve

Gene Therapy Comes of Age

Within two months of his surgery, Jack's visual improvement was remarkable. He exulted, "I don't have to hold onto my friend's shoulders anymore when I go to a movie theater or go outside at night. I've never ridden my bike at night, and now I can."[43] No one was sure that the one-time gene therapy would last, but a year after his treatment, Jack continued to improve. Other patients who have received the therapy continue to maintain their improved vision even after four years. Gene therapy with Luxturna is so new that researchers do not yet know whether the results are permanent; however, surgical eye specialists Kenneth C. Fan and Audina M. Berrocal say, "Gene therapy for [inherited retinal diseases] is ushering the retinal physician into an era that was nearly unimaginable just a decade ago."[44]

Currently, at least nine gene therapies have received approval in the United States and Europe for use with certain kinds of cancers, infections, and inherited diseases. Science journalist Jim Daley says, "After nearly half a century, the concept of genetic medicine has become a reality."[45] One approved therapy is for melanoma, a kind of skin cancer. The therapy, known as Imlygic, entails inserting a gene into the cancer to code for killing the cancer cells. Another, called Strimvelis, treats severe immune

The High Cost of Gene Therapy

Gene therapies offer almost miraculous benefits to people with genetic diseases, but the cost for these breakthrough treatments is stunning. Zolgensma, for example, the therapy for the treatment of spinal muscular atrophy (SMA) costs $2.125 million for the one-time treatment. Critics express outrage at what they see as excessive profit for the drug company, but Nathan Yates disagrees. Yates is an economics and finance professor who also has SMA. His is a milder form of the disease that did not take his life, but he is disabled and wheelchair bound. Instead of being outraged by the price, Yates is thrilled by the potential of the new therapy. He says he is "perplexed and disappointed" by the criticism. He explains, "How are we going to get treatments for rare diseases if there's not a financial incentive for doing it? . . . Don't we realize, though, that all of society profits from each disease we cure and each baby that is saved from SMA and other deadly diseases?" He points out that the price is perhaps not so bad when you consider that the cost of renting a ventilator to breathe is $1,000 per month. He adds, "We should not put a price tag on life. . . . Instead, think about the parents who will no longer have to receive the heartbreaking news that my parents were given 29 years ago: 'Your child has spinal muscular atrophy, and there's nothing we can do. Survival beyond early childhood is unlikely.' The price of Zolgensma seems insignificant now, don't you think?"

Nathan Yates, "I Have Spinal Muscular Atrophy. Critics of the $2 Million New Gene Therapy Are Missing the Point," STAT, May 31, 2019. www.statnews.com.

deficiency disease by correcting the lack of white blood cells in the immune system. Still another, Trikafta was approved in 2019 and is a treatment for most forms of cystic fibrosis, a serious, life-threatening genetic disease. Trikafta is expected to be life-changing for most people with cystic fibrosis.

No Viral Vectors Needed

The biotechnology company that developed Trikafta, named Vertex, is working on experimental gene therapies for other diseases. It has partnered with another company, CRISPR Therapeutics, to research gene-editing technology to treat such diseases as sickle-cell disease, a painful red blood cell disorder, and beta thalassemia, a sometimes deadly blood disorder that reduces the oxygen supply to the body. CRISPR technology would not require a viral vector. Instead, body cells could be removed from the malfunctioning organ, corrected in the lab, and then put back into the body.

The gene-edited cells could then provide the proper coding instructions to cure the disease. Some researchers are also researching how to use CRISPR to make gene edits in organs inside the body. CRISPR might make an enormous difference for people suffering with the more than six thousand known genetic diseases. Science writer MaryAnn Labant says, "Genome-editing tools open a wide range of new possibilities in gene manipulation including target-specific gene repair."[46]

CRISPR Research

Gene editing with CRISPR is still in the research stage, but results are encouraging. Researchers at Exonics Therapeutics have reported successful gene editing of Cavalier King Charles spaniels that have Duchenne muscular dystrophy. In people, this progressive disease leads to increasing weakness of muscles that is eventually fatal. It is caused by a mutation in the dystrophin gene. In the dogs, the researchers used CRISPR to correct the dystrophin gene and were able to restore the dogs' limb and heart muscle function. The technique has not yet been tried in humans, but Vertex bought Exonics Therapeutics in 2019 and plans to aggressively expand the research. The chairman and chief executive officer of Vertex, Jeffrey Leiden, says, "Our single focus at Vertex is to bring transformative medicines to people with serious diseases including Duchenne."[47]

A company called Editas Medicine is beginning a gene-editing clinical trial to treat people with an inherited kind of blindness known as Leber congenital amaurosis 10. This disease is caused by a malfunctioning CEP290 gene. Unlike the RPE65

Researchers have reported successful gene editing of Cavalier King Charles spaniels (pictured) that have Duchenne muscular dystrophy. They were able to restore the dogs' limb and heart muscle function.

gene involved in retinitis pigmentosa, the CEP290 gene is too big to insert in a viral vector. First the company must test very low doses of their gene-editing CRISPR therapy by inserting it directly into the retinas of a few volunteers who are almost completely blind. If all goes well, they will expand the tests. Charles Albright, chief scientist of Editas Medicine, explains, "We're going into arguably the most difficult patients to start with and we're going to improve from there."[48]

Genetic researchers are extremely hopeful and excited about the potential of CRISPR gene therapy to solve previously unsolvable problems. Laurie Zoloth, a bioethicist at the University of Chicago Divinity School, says, "I want it to work. Everyone who thinks seriously about human suffering should really be wanting this to happen and should be optimistic . . . about medicine's capacity and its power."[49]

A Genetic Engineering Nightmare?

After her discovery of CRISPR technology, Jennifer Doudna had a nightmare. She recounts, "So, one night, I dreamed that a colleague said, 'I need to introduce you to someone.' This man turned around and it was *Hitler*. But you know, the dream really brought home to me the profound nature of what we were doing." In Doudna's nightmare, Hitler wanted to learn about CRISPR and how he could use it to genetically modify the human race. She was horrified. Would her discovery be used for evil? Hitler believed in eugenics, which is the theory that a superior race of humans can be created through genetic manipulation. Doudna explains that CRISPR enables "cures for genetic diseases and conditions, an increased food supply. But it also brings the potential for eugenics, for state-sponsored alteration of human beings. You even can imagine creating new species of humans." She also knows that, in the future, it could be used for creating so-called designer babies just because the parents have the money and desire to do so. She wonders whether the day will come when people will gene-edit embryos to make the children have blue eyes, be six feet tall, or have high intelligence, and whether humanity really wants to embrace such scenarios. She urges scientists and all people to begin thinking about how to use gene-editing technology ethically. She concludes, "I don't have any regrets about the science we've done. [But] I think that I would regret not being vocally involved in advocating for the responsible use of it."

Quoted in Claudia Dreifus, "'The Joy of the Discovery': An Interview with Jennifer Doudna," NYR Daily, January 24, 2019. www.nybooks.com.

When Should CRISPR Be Used?

Bioethicists study the ethics, or moral issues, that can arise in the fields of biology and medicine. Few ethical issues arise when scientists attempt to cure diseases with gene therapy, but all of the gene therapies so far involve modification of somatic, or body, cells. Somatic cells make up the various tissues of the body, and when their genes are edited, the change affects that specific kind

somatic
Referring to the body

of cell only. Germline cells, on the other hand, are reproductive cells and do present ethical questions. They carry all the genetic information to create a new human being. Bioethicists Sarah Polcz and Anna Lewis explain: "Germline cells are those that have the potential to be inherited by the next generation; for example, eggs and sperm. Changes that are introduced to germline cells enter the gene pool of that species."[50]

germline
Referring to the reproductive cells such as eggs and sperm

When scientists edit the genes of plants and animals, they are creating changes that are passed down to future generations that carry that altered DNA. The obvious question is whether this is something researchers should do with people. Should researchers attempt to change an embryo to create humans with altered DNA to fight disease or improve the species? Almost to a person, scientists and bioethicists say no. Not only has the scientific community agreed that such experiments are unwise and too dangerous for now, the World Health Organization (WHO) recommends that the world's governments regulate such research.

gene pool
The total numwber of genes of every individual in a population

A Major Ethical Concern

Nevertheless, in 2018 Chinese scientist He Jiankui announced that he had gene-edited twin embryos and implanted them in their mother. Two healthy girls were born. Most scientists were shocked and outraged. He claimed to have edited the embryos to prevent them from developing HIV, which infected their father, but other scientists insisted the procedure was unnecessary. HIV is only rarely passed to offspring, and no one knows whether there will be dangerous long-term effects to these children.

He was arrested and jailed in China for his experiment. Jennifer Doudna explains: "The repair was different in different embryos, so he created changes to the DNA that honestly had probably never been seen in the human population and never even tested in animals." She says such research is irresponsible and cannot be tolerated. Still, she knows germline gene-editing experiments will be done again, and she wants everyone to begin thinking about how humanity will make rules about the uses of gene editing in the future. She adds, "I would say my feeling today is that, like it or not, we're going to have to figure it out."[51] Genetic engineering holds great promise but also perhaps some peril for the future of humanity.

SOURCE NOTES

Introduction: Changed Forever
1. Quoted in Alvin Powell, "Gene Therapy Was a 'Last Shot,'" *Harvard Gazette*, February 21, 2019. https://news.harvard.edu.
2. Quoted in Powell, "Gene Therapy Was a 'Last Shot.'"
3. Bryan Walsh, "How Should Genetic Engineering Shape Our Future?," LeapsMag Newsletter, April 27, 2018. https://leapsmag.com.

Chapter One: What Is Genetic Engineering?
4. Quoted in Clara Rodríguez Fernández, "10 Unusual Applications of CRISPR Gene Editing," Labiotech.eu, April 3, 2019. www.labiotech.eu.
5. National Human Genome Research Institute, "DNA Sequencing Fact Sheet," National Institutes of Health, December 18, 2015. www.genome.gov.
6. Edd Gent, "With These 4 Breakthroughs, We'll Be Able to Write Whole Genomes from Scratch," Singularity Hub, October 21, 2019. https://singularityhub.com.
7. Jon Entine and XiaoZhi Lim, "Cheese: The GMO Food Die-Hard GMO Opponents Love, but Don't Want to Label," Genetic Literacy Project, November 2, 2018. https://geneticliteracyproject.org.
8. Quoted in Alvin Powell, "CRISPR's Breakthrough Implications," *Harvard Gazette*, May 16, 2018. https://news.harvard.edu.
9. Jennifer Doudna, "The Gene-Editing Revolution Is Already Here," *Time*, October 24, 2019. https://time.com.

Chapter Two: Drugs and Vaccines for Fighting Disease
10. Quoted in Rachael Rettner, "New HIV Vaccine Study Starts in South Africa," Live Science, November 28, 2016. www.livescience.com.
11. Quoted in Amy Schleunes, "Another HIV Vaccine Clinical Trial Fails," *The Scientist*, February 3, 2020. www.the-scientist.com.

12. Quoted in Schleunes, "Another HIV Vaccine Clinical Trial Fails."
13. Kris Heeter, "Recombinant DNA Technology for Vaccine Development," Sciencing, September 17, 2018. https://sciencing.com.
14. Centers for Disease Control and Prevention, "How Influenza (Flu) Vaccines Are Made," December 12, 2019. www.cdc.gov.
15. Heeter, "Recombinant DNA Technology for Vaccine Development."
16. Seyed Davoud Jazayeri and Chit Laa Poh, "Recent Advances in Delivery of Veterinary DNA Vaccines Against Avian Pathogens," *Veterinary Research*, vol. 50, no. 78, October 10, 2019, p.10. https://veterinaryresearch.biomedcentral.com.

Chapter Three: Engineering Food Plants for People

17. Connor McKoy, "Gene Editing Can Help Agriculture Be Climate-Friendly . . . Here's How," *BIOtechNow* (blog), September 9, 2019. www.bio.org.
18. Quoted in McKoy, "Gene Editing Can Help Agriculture Be Climate-Friendly . . . Here's How."
19. Quoted in Jonathan Knutson, "The Benefits of Bt Corn—Study Finds Broader Gains," *Agweek*, March 21, 2018. www.agweek.com.
20. Quoted in Marc Lallanilla, "What Are GMOs and GM Foods?," Live Science, July 8, 2019. www.livescience.com.
21. US Food and Drug Administration, "Feed Your Mind," Agricultural Biotechnology, March 4, 2020. www.fda.gov.
22. US Department of Agriculture, "BE Disclosure." www.ams.usda.gov.
23. Quoted in Joan Conrow, "Developing Nations Lead Growth of GMO Crops," Cornell Alliance for Science, June 29, 2018. https://allianceforscience.cornell.edu.
24. Quoted in Kristin Houser, "CRISPR Co-inventor: We'll Be Eating Gene-Edited Food in Five Years," Futurism, April 22, 2019. https://futurism.com.
25. Quoted in Chuck Gill, "Penn State Developer of Gene-Edited Mushroom Wins 'Best of What's New' Award," Penn State News, October 19, 2016. https://news.psu.edu.

26. Jennifer Doudna, "Berkeley Talks Transcript: Jennifer Doudna on the Future of Gene Editing," Public Affairs, Berkeley News, April 10, 2019. https://news.berkeley.edu.
27. Quoted in Jon Cohen, "To Feed Its 1.4 Billion, China Bets Big on Genome Editing of Crops," *Science*, July 29, 2019. www.sciencemag.org.
28. Quoted in Megan Molteni, "The First Gene-Edited Food Is Now Being Served," *Wired*, March 20, 2019. www.wired.com.
29. Calyxt. https://calyxt.com.

Chapter Four: Food Animals and Genetic Engineering

30. Quoted in Amy Nordrum, "U.S. Consumers Might Get Their First Taste of Transgenic Salmon This Year," *IEEE Spectrum*, January 5, 2020. https://spectrum.ieee.org.
31. US Food and Drug Administration, "Q&A on FDA Regulation of Intentionally Altered Genomic DNA in Animals," October 13, 2017. www.fda.gov.
32. Quoted in Cameron English, "FDA Defends CRISPR-Edited Animal Rules Likely to Block Most Uses: Is the Agency Trying to Avoid Litigation from Anti-GMO Groups?," Genetic Literacy Project, February 11, 2020. https://geneticliteracyproject.org.
33. Quoted in Jon Cohen, "China's CRISPR Push in Animals Promises Better Meat, Novel Therapies, and Pig Organs for People," *Science*, July 31, 2019. www.sciencemag.org.
34. Quoted in Cohen, "China's CRISPR Push in Animals Promises Better Meat, Novel Therapies, and Pig Organs for People."
35. Quoted in Pallab Ghosh, "Gene-Edited Animal Plan to Relieve Poverty in Africa," BBC News, February 15, 2019. www.bbc.com.
36. Quoted in Ghosh, "Gene-Edited Animal Plan to Relieve Poverty in Africa."
37. Quoted in GEN: Genetic Engineering & Biotechnology News, "CRISPR Chicken Cells Have Secret Recipe to Resist Bird Flu," June 5, 2019. www.genengnews.com.
38. Quoted in GEN: Genetic Engineering & Biotechnology News, "CRISPR Chicken Cells Have Secret Recipe to Resist Bird Flu."

39. Quoted in Carolyn Y. Johnson, "Gene-Edited Farm Animals Are Coming. Will We Eat Them?," *Washington Post*, December 17, 2018. www.washingtonpost.com.
40. Christine Tait-Burkard et al., "Livestock 2.0—Genome Editing for Fitter, Healthier, and More Productive Farmed Animals," *Genome Biology*, vol. 19, no. 204, November 26, 2018. www.ncbi.nlm.nih.gov.

Chapter Five: Altering Human Genes

41. Cleveland Clinic, "Spinal Muscular Atrophy (SMA)," November 24, 2015. https://my.clevelandclinic.org.
42. Quoted in Children's Hospital of Philadelphia, "Gene Therapy Treatment for Spinal Muscular Atrophy: Céline's Story," August 23, 2019. www.chop.edu.
43. Quoted in Beacon Stories, "Jack Sees a Different Life After LUXTURNA," Foundation Fighting Blindness, June 17, 2019. www.fightingblindness.org.
44. Kenneth C. Fan and Audina M. Berrocal, "Surgical Techniques for Retinal Gene Therapy Delivery," *Retinal Physician*, February 1, 2020. www.retinalphysician.com.
45. Jim Daley, "Gene Therapy Arrives," *Scientific American*, January 1, 2020. www.scientificamerican.com.
46. MaryAnn Labant, "CRISPR Jump-Starts Gene Therapy," GEN: Genetic Engineering & Biotechnology News, April 1, 2019. www.genengnews.com.
47. Quoted in Business Wire, "Nonprofit CureDuchenne Is Encouraged to See Gene Editing for Duchenne Advance Through Vertex's Acquisition of Exonics Therapeutics," June 6, 2019. www.businesswire.com.
48. Quoted in Tina Hesman Saey, "CRISPR Enters Its First Human Clinical Trials," *ScienceNews*, August 14, 2019. www.sciencenews.org.
49. Quoted in Saey, "CRISPR Enters Its First Human Clinical Trials."
50. Sarah Polcz and Anna Lewis, "CRISPR-Cas9 and the Nongermline Non-controversy," *Journal of Law and the Biosciences*, vol. 3, no. 2, August 2016, p. 413. https://academic.oup.com.
51. Quoted in Hannah Kuchler, "Jennifer Doudna, Crispr Scientist, on the Ethics of Editing Humans," *Financial Times*, January 31, 2020. www.ft.com.

FOR FURTHER RESEARCH

Books
Tara Haelle, *Vaccination Investigation: The History and Science of Vaccines*. Minneapolis: Lerner, 2018.

Susan Henneberg, *Genetic Engineering*. New York: Greenhaven, 2017.

Tom Jackson, *What's the Issue: Future Humans*. London: QED, 2019.

Don Nardo, *How Vaccines Changed the World*. San Diego: ReferencePoint, 2018.

Yolanda Ridge, *CRISPR: A Powerful Way to Change DNA*. Toronto: Annick Press, 2020.

Internet Sources
Heidi Ledford, "Gene Therapy Is Facing Its Biggest Challenge Yet," *Nature*, December 4, 2019. www.nature.com.

Linda Marsa, "Gene Therapies Make It to Clinical Trials," *Discover*, December 31, 2019. www.discovermagazine.com.

ProCon.org, "GMOs - Top 3 Pros and Cons," March 10, 2020. www.procon.org.

Tina Hesman Saey, "Explainer: How CRISPR Works," *Science News for Students*, July 31, 2017. www.sciencenewsforstudents.org.

Websites
American Society of Gene and Cell Therapy (www.asgct.org). Learn the basics of gene therapy at this professional site and keep abreast of the latest gene therapy news. Visitors can click the Education link to explore the research for many different rare genetic diseases.

Genetic Literacy Project (https://geneticliteracyproject.org). This nonprofit group encourages "Science Not Ideology" and is basically a library of information and articles explaining the facts of genetics and biotechnology topics from agriculture to animals to human gene therapy. Click on topics of interest on the Home page or search for any genetics topic.

Genetics Generation (https://knowgenetics.org/). At this site, visitors can learn genetics basics, test their knowledge of genetics, read about people with genetic issues, and explore articles about the ethics of genetic alterations.

yourgenome.org (www.yourgenome.org). With a variety of articles, animations, and videos, this website explores a wide assortment of topics related to genes, DNA, and the implications of the science of genetic technology for society and individual health.

INDEX

Note: Boldface page numbers indicate illustrations.

Acceligen, 47–48
acute lymphoblastic leukemia (ALL), **59**
  gene therapy for, 58–59
adenine, 11
African Swine Fever (ASF), victims of, **52**
American Association for the Advancement of Science (AAAS), 37
American Medical Association, 37
American Society of Gene and Cell Therapy, 71
antibody, definition of, 28
antigen, definition of, 28
AquaBounty Technologies, 44–45, 45–46

Bacillus thuringiensis (Bt), 34
bacteria, 23–24
  cell division in, 12
  immune system's response to, 28
  production of insulin with, 24–26, **27**
  restriction enzymes and, 15
  vaccines made of, 28
  as vectors, 34

bananas, genetically modified, 42
bases, of DNA molecule, 10–11, 19
Berrocal, Audina M., 60
Bill and Melinda Gates Foundation, 48
bioengineered, definition of, 38
blood cells, 7
bone marrow, 7
Boyer, Herb, 23, 24–25
British Royal Society, 37
Bwesigye, Priva Namanya, 42

Calyno (modified soybean oil), 43
cattle, heat-resistant, 47–48
Cavalier King Charles spaniel, 62, **63**
cell division, 24
  errors in, 12–13, 19
cell-mediated immunity, 31
Centers for Disease Control and Prevention (CDC), 29
Charpentier, Emmanuelle, 16–17, 18
chickens, Newcastle disease-resistant, 48–50
chocolate, use of CRISPR to save from climate change, 36
chromosome(s)

73

definition of, 12
number of, in humans, 12
chronic granulomatous
   disease (CGD), 6, 7
Clustered Regularly
   Interspaced Short
   Palindromic Repeats
   (CRISPR), 16–18
cocoa swollen shoot virus, 36
coffee, genetic engineering of,
   10
Cohen, Stanley, 23, 24–25
corn
   genetically modified as
      percentage of total grown
      in US, 37
   genetically modified Bt,
      34–35
COVID-19 (coronavirus
   disease)
   virus causing, **30**
   work on vaccine for, 25
Crisanti, Andrea, 50
CRISPR/Cas9, 16–18, 33–34
   ethical issues concerning,
      64–66
   to fight avian influenza,
      50–51
   in modifying cacao to
      withstand climate change,
      36
   in removing allergen from
      food, 42–43
   use in food crops, 39–40
CRISPR Therapeutics, 61–62

cytosine, 11

Daley, Jim, 60
Dana-Farber/Boston Children's
   Cancer and Blood Disorders
   Center, 7
deoxyribonucleic acid (DNA), 8
   genetic engineering and, 10
   genome sequencing and, 14
   mutations in, 12–13, 19
   RNA *vs.*, 17–18
   structure of, 10–11
   See also recombinant DNA
      technology
Department of Agriculture, US
   (USDA), 32, 38, 39–40
Djikeng, Appolinaire, 48
DNA. *See* deoxyribonucleic
   acid
double helix, 11
Doudna, Jennifer, 16–17, 18,
   18, 19–20
   on Chinese government
      support of CRISPR
      research, 46–47
   on CRISPR technology and
      food supply, 39, 40–41
   on ethical concerns with
      CRISPR technology, 64
   on germline gene-editing
      experiments, 65–66
drugs, recombinant DNA, 26
   insulin, 23–26
Duchenne muscular dystrophy,
   62

dwarfism, 26

Editas Medicine, 62–63
eugenics, 64

Fan, Kenneth C., 60
Fauci, Anthony, 21, 22
Feng Zhang, 18
food, bioengineered, 38–39
Food and Drug Administration, US (FDA), 44, 54
   on food wasted in US, 40
   on GM food, 37–38
   ruling on gene editing in food animals, 45
food waste, CRISPR technology and, 40–41

gamete(s), definition of, 12
Gao Caixia, 41–42
gene editing, 16–19
   genetic modification *vs.*, 34
   See also CRISPR/Cas9
gene pool, definition of, 65
gene(s), 11
   definition of, 7
   number of, in humans, 12
   as units of DNA, 11, **11**
gene therapy
   for acute lymphoblastic leukemia, 58–59
   cost of, 61
   for retinitis pigmentosa, 59–60
   for spinal muscular atrophy, 55–57, 61
genetically modified organisms (GMOs), 34
   controversy over, 35–37
   labeling of GM food, 38
   made with recombinant DNA *vs.* CRISPR technology, 39–40
genetic engineering, first success in, 22–25
Genetic Literacy Project, 72
Genetics Generation, 72
genome
   complexity in sequencing of, 13
   definition of, 13
   human, size of, 13–14
genome sequence, 13
Gent, Edd, 14
germline, definition of, 65
germline cells, 65
Gershon, Gilad, 10
GMOs. *See* genetically modified organisms
granulomas, 6
Gray, Glenda, 22
Grubor-Bauk, Branka, 31
guanine, 11
Guiltinan, Mark, 36

Haemophilus influenzae, 13
Hallerman, Eric, 45
Heeter, Kris, 29, 30
He Jiankui, 65–66

hemagglutinin (HA), 29
hemophilia, 26
hepatitis B vaccine, 29
Hitler, Adolf, 64
HIV vaccine, South African trial of, 21–22, **23**
Hogan, Jack, 59–60
human growth hormone (HGH), 26
hydrogenated, definition of, 43

Imlygic (gene therapy), 60
immune system, 28
   definition of, 6
influenza, avian, CRISPR technology to fight, 50–51
influenza vaccine
   DNA, for H5N1 virus in chickens, 29
   recombinant, 29
inheritance, 12–13
insulin, 22–23, 26
   recombinant human, creation of, 23–25, **27**
Intergovernmental Panel on Climate Change (United Nations), 33
International Service for the Acquisition of Agri-biotech Applications (ISAAA), 39
intron(s), 24
   definition of, 24

Jazayeri, Seyed Davoud, 32

Kichula, Elizabeth, 56
Kuzma, Jennifer, 54
Kymriah (gene therapy), 58–59

Labant, MaryAnn, 62
Lai Liangxue, 52–53
Leber congenital amaurosis 10, 62–63
Leiden, Jeffrey, 62
Lewis, Anna, 65
Lippman, Zach, 40–41
Luxturna (gene therapy), 60

McKoy, Connor, 33
medicine, genetic. *See* gene therapy
messenger RNA (mRNA), 25
mosquitoes, gene editing to eliminate, 50
mushrooms, 40, **41**
mutation(s), 8, 12
   average number of, at birth, 13
   how diseases are caused by, 19
   off-target, 17

National Academy of Sciences, 37
National Agricultural Research Laboratories (NaRL, Uganda), 42
National Human Genome Research Institute (National Institutes of Health), 13

National Institute of Allergy and Infectious Diseases (NIAID), 21
National Institutes of Health (NIH), 13
neutrophils, 6, 7

off-target mutations, 17
organ transplants, 53

pancreas, 22
pegRNA, 17
pigs
   CRISPR technology to fight African Swine Fever in, 51–54
   as source of transplant organs, 53
pituitary glands, 26
plasmid(s), 24
   definition of, 24
   DNA vaccines and, 32
Poh, Chit Laa, 32
Polcz, Sarah, 65
Prather, Randall, 46
prime editing, 17
proteins
   genes carry code for making, 11
   role of, 12

Qi, Yiping, 34

recombinant DNA technology, 15–16
in genetically modified food crops, 34–39
insulin produced by, 22–25
other drugs produced by, 26
reservoir, definition of, 51
resilience, definition of, 53
restriction enzymes, 15–16, 24
retina, definition of, 60
retinitis pigmentosa, 59–60
ribonucleic acid (RNA)
   messenger, 25
   role of, 17–18

salmon, **46**
   gene-edited, 44–45
SARS-CoV-2, 25, **30**
sexual reproduction, 12
sickle-cell disease, 62
Solomon, Steven, 45
somatic, definition of, 64
somatic cells, 64
Sonstegard, Tad, 47–48
spinal muscular atrophy (SMA), 55, 61
Spinraza (gene therapy), 55, 56
Strimvelis (gene therapy), 60–61
subsistence farming, definition of, 48

Tait-Burkard, Christine, 54
TALENS (molecular editing technology), 54
Teng, Paul S., 39

thymine, 11
Tropic Biosciences, 10

University of Maryland, 36

vaccines, 21
  for COVID-19, 25
  DNA (third generation), 30–32
  immune system and, 28
  from killed/weakened viruses (first generation), 28–29
  recombinant DNA (second generation), 29
  for Zika virus, 31
vector, definition of, 34
Vertex, 61–62
virus(es)
  causing COVID-19, 25, **30**
  immune system's response to, 28
  surface proteins on, 29
  in vaccines, 28–29
  as vectors, 34

Zika, work on vaccine for, 31
Voytas, Dan, 43

Walsh, Bryan, 9
Wang Haoyi, 47
Warren, Mitchell, 22
Whittaker, Brenden, 6–7, 8
World Health Organization (WHO), 37, 65

xenotransplantation, 53

Yang, Yinong, 40
Yates, Nathan, 61
yourgenome.org, 72

Zhou Qi, 47
Zika virus, breakthrough in search for vaccine for, 31
Zolgensma (gene therapy), 56–57
  cost of, 61
Zoloth, Laurie, 63

PICTURE CREDITS

Cover: vectorfusionart/Shutterstock.com

8: science photo/Shutterstock.com
11: ktsdesign/Shutterstock.com
14: oneinchpunch/Shutterstock.com
18: Associated Press
23: Avatar_023/Shutterstock.com
27: Emre Terim/Shutterstock.com
30: Gemini Pro Studio/Shutterstock.com
35: Stephen William Robinson/Shutterstock.com
38: Vladimir Mulder/Shutterstock.com
41: Picture Partners/Shutterstock.com
46: oceanfishing/Shutterstock.com
49: Nick Fox/Shutterstock.com
52: galitsin/Shutterstock.com
57: Suzanne Grala/Science Source
59: Kateryna Kon/Shutterstock.com
63: Alexandra Morrison Photo/Shutterstock.com

ABOUT THE AUTHOR

Toney Allman holds degrees from Ohio State University and the University of Hawaii. She currently lives in Virginia, where she enjoys a rural lifestyle, as well as researching and writing about a variety of topics for students.